STUDENT SUPPLEMENT

Volume 1: Chapters 1-12

to accompany

Swokowski's CALCULUS With Analytic Geometry
Second Edition

by

Thomas A. Bronikowski
Marquette University

Prindle, Weber & Schmidt
Boston, Massachusetts

© Copyright 1979 by PRINDLE, WEBER & SCHMIDT
20 Providence Street, Boston, Massachusetts 02116

ISBN 0-87150-277-1

Third printing: August, 1980

Prindle, Weber & Schmidt is a division of Wadsworth, Inc.

INTRODUCTION

This supplement to the second edition of Earl W. Swokowski's text, <u>Calculus with Analytic Geometry</u>, has been written to help you, the student, develop the skills and art of solving Calculus problems. In it I have included the solution of every third problem (numbers 1, 4, 7, 10, etc.) from every section of the text. Very occasionally this scheme is slightly modified to provide a greater variety of exercises and to furnish hints for some problems not in the usual sequence. There is a special feature of this second edition which should be of significant help to you in sections of the text concerned with applications of the integral. (These are principally in Chapters 6 and 10 of Volume 1 and Chapters 13, 15 and 17 of Volume 2.) In addition to the complete solutions of the usual exercises, I have included in such sections the integral forms of the answers to the remaining odd-numbered exercises (numbers 3, 5, 9, 11, etc.). In this way you can check the "set-up" of the solution of the exercise before wasting precious time evaluating the wrong integral.

Obviously not every arithmetic and algebraic detail can be included in every solution if the size of this work is to remain reasonable. You will find sufficient detail in every solution to see how the answer is obtained. As a rule, more calculational details are shown in the first problem of a set than in the last problems of that set. The symbols " $\Rightarrow$ " and " $\Leftrightarrow$ " are frequently used to denote implication and equivalence, respectively, between the mathematical expressions they connect. All other symbols will be familiar to you from the text or from previous courses. The author would appreciate being informed of any errors or misprints that you find. Just send a note to me at the Department of Mathematics and Statistics, Marquette University, Milwaukee, WI, 53233.

I wish to thank all of those who have spoken or written to me pointing out errors and suggesting improvements in the first edition. I would also like to express my appreciation to the staff of Prindle, Weber and Schmidt, especially John Martindale and David Chelton, for their assistance and advice, to my friend and colleague, Professor Earl W. Swokowski, for his thoughtful recommendations, and to Sally Canapa for her excellent job of typing the entire manuscript. To my wife, Irene, and our children, Joseph, Michael and Anne, I owe a special debt of gratitude for their encouragement and assistance in the preparation of this edition. Particular appreciation is expressed to the boys for their help in proofreading and in making several new sketches and to my wife for her generous and unselfish assistance in proofreading and in preparing the final version of the finished manuscript.

Thomas A. Bronikowski
Department of Mathematics and
   Statistics
Marquette University
Milwaukee, WI 53233

TABLE OF CONTENTS

EXERCISES 1.1, page 9

1. (a) $-2 > -5$ since $-2 - (-5) = 3 > 0.$ (b) $-2 < 5$ since $5 - (-2) = 7 > 0.$
    (c) $(6-1) = (2+3)$ since $(6-1) - (2+3) = 0.$ (d) $2/3 > 0.66$ since $2/3 - 0.66 =$
    $0.666... - 0.66 = 0.006... > 0.$ (e) $2 = \sqrt{4}$ by definition of the $\sqrt{}$ symbol.
    (f) $\pi < 22/7$ since $22/7 - \pi = 3.14285... - 3.14159... = 0.00126...> 0.$

4. (a) $|4-8| = |-4| = -(-4) = 4.$ (b) $|3-\pi| = -(3-\pi) = \pi-3.$ (c) $|-4| - |-8| =$
    $-(-4) - (-(-8)) = 4-8 = -4.$ (d) $|-4+8| = |4| = 4.$ (e) $|-3|^2 = (-(-3))^2 = 3^2$
    $= 9.$ (f) $|2-\sqrt{4}| = |2-2| = |0| = 0.$ (g) $|-0.67| = -(-0.67) = 0.67.$
    (h) $-|-3| = -[-(-3)] = -3.$

7. $5x-6 > 11 \iff 5x > 17 \iff x > 17/5.$ $\therefore (17/5,\infty)$ is the solution set.

10. $7-2x \geq -3 \iff 10 \geq 2x \iff 5 \geq x.$ $\therefore (-\infty,5]$ is the solution set.

13. $3x+2 < 5x-8 \iff 10 < 2x \iff 5 < x.$ $\therefore (5,\infty)$ is the solution set.

16. $-4 <. 2-9x < 5 \iff -6 <-9x < 3 \iff 2/3 > x > -1/3.$ $\therefore (-1/3,2/3)$ is the
    solution set.

19. $5/(7-2x) > 0 \iff 7-2x > 0$ (since $5 > 0$) $\iff 7 > 2x \iff 7/2 > x.$
    $\therefore (-\infty,7/2)$ is the solution set.

22. $|(2x+3)/5| < 2 \iff |2x+3| < 10 \iff -10 < 2x+3 < 10 \iff -13 < 2x < 7 \iff$
    $-13/2 < x < 7/2$ or $(-13/2,7/2).$

25. $|25x-8| > 7 \iff 25x-8 > 7$ or $25x-8 < -7.$ The first yields $x > 3/5;$ the
    second yields $x < 1/25.$ Thus the solution set is $(-\infty,1/25) \cup (3/5,\infty).$

28. $2x^2 - 9x + 7 < 0 \iff (2x-7)(x-1) < 0.$ For the product to be negative, the
    factors must differ in sign. The points where one or the other changes sign
    are $x = 1$ and $x = 7/2.$ Schematically we have

    ```
    (2x-7)      -           -         +
            ─────────────┼─────────┼─────── x
    (x-1)       -       1   +    7/2  +
    ```

    Thus $(1, 7/2)$ is the desired solution set.

31. $1/x^2 < 100 \iff 1/|x| < 10 \iff |x| > 1/10 \iff x > 1/10$ or $x < -1/10.$ Thus
    $(-\infty,-1/10) \cup (1/10,\infty)$ is the solution set.

34. It is convenient to look at the inequality separately on $(-\infty,-2)$, $(-2,9)$,
    $(9,\infty)$. On the 1st, both denominators are negative and $3/(x-9) > 2/(x+2) \iff$
    $(3x+6)/(x-9) < 2 \iff 3x+6 > 2x-18 \iff x > -24.$ Since $x \in (-\infty,-2)$, we get
    $(-24,-2).$ On the 2nd interval the left side is < 0, the right side is > 0,
    and, so, no number here can satisfy the given inequality. On the 3rd, $(9,\infty)$,
    both denominators are positive and, as above, we obtain $x > -24.$ Since every
    $x \in (9,\infty)$ satisfies this, we get the entire interval $(9,\infty)$. Combining

results we get $(-24,-2) \cup (9,\infty)$ as the solution set.

37. Hint. Multiply $0 < a < b$ by $1/a > 0$ and then by $1/b > 0$. If $a = 0$, $1/a$ does not exist. If $a < 0$, then $1/a < 0 < 1/b$.

40. No. The result is true if all numbers are positive, but the reverse in-equality is true if all are negative. In mixed cases, anything can happen.

EXERCISES 1.2, page 19

1. (a) $d(A,B) = \sqrt{(2-6)^2 + (1-(-2))^2} = \sqrt{(-4)^2 + 3^2} = \sqrt{16+9} = \sqrt{25} = 5$. (b) Midpoint is $((6+2)/2,(-2+1)/2) = (4,-1/2)$.

4. (a) $d(A,B) = \sqrt{(4-4)^2 + (5-(-4))^2} = \sqrt{0+9^2} = 9$. (b) Midpoint is $((4+4)/2,(5-4)/2) = (4,1/2)$.

7. Since $d(A,B)^2 = 5^2 + 5^2 = 50$, $d(B,C)^2 = 7^2 + 7^2 = 98$, $d(A,C)^2 = 12^2 + 2^2 = 148$, it is a right triangle with legs AB and BC and hypotenuse AC. The area is $\frac{1}{2}d(A,B)d(B,C) = \frac{1}{2}\sqrt{50}\sqrt{98} = (5\sqrt{2})(7\sqrt{2})/2 = 35$.

10. A point (x,y) satisfies $y = -3$ if and only if it is 3 units below the x-axis. Thus the graph is a horizontal line 3 units below and parallel to the x-axis.

13. $|x| < 2 \iff -2 < x < 2$ and $|y| > 1 \iff y > 1$ or $y < -1$. Thus the graph consists of the two vertical strips shown between the vertical lines $x = \pm 2$. The upper strip corresponds to $y > 1$; the lower to $y < -1$.

NOTE: In graphing the equations in #15-30 of the text, all you can do now is tabulate some points, plot them and connect them smoothly. Only after Chapters 3, 4 and 7 will you have the mathematical background to sketch graphs accurately with only a few significant points plotted. The graphs for the odd-numbered exercises are in the text.

16. For $y = 4x-3$, we obtain the table of points and then plot to obtain the line shown.

x	-2	-1	0	1	2
y	-11	-7	-3	1	5

19. For $y = 2x^2 - 1$, we plot

x	± 2	± 1	0
y	7	1	-1

to obtain the graph in the text, a parabola.

22. Rewrite the equation as $y = -x^2/3$ and tabulate as above.

x	±3	±2	±1	0
y	-3	-4/3	-1/3	0

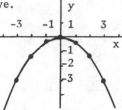

25. For $y = x^3 - 2$, we obtain

x	-2	-1	0	1	2
y	-10	-3	-2	-1	6

28. For $y = \sqrt{x} - 1$, x must be ≥ 0 and we get

x	0	1	4	9
y	-1	0	1	2

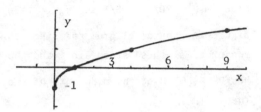

31. By (1.9), the graph is a circle of radius 4 with center at (0,0).

34. Directly from (1.9) the equation is $(x-(-5))^2 + (y-2)^2 = 5^2$ or $(x+5)^2 + (y-2)^2 = 25$.

37. Since the center has y-coordinate 2, the radius must be 2 for the circle to be tangent to the x-axis. Thus the equation is $(x-(-4))^2 + (y-2)^2 = 2^2$, which reduces to the answer given.

40. The center must be 2 units from each axis for the circle to have radius 2 and to be tangent to both axes. Since it is in the 1st quadrant, the center is at (2,2) and $(x-2)^2 + (y-2)^2 = 2^2$ is the equation.

43. We solve by completing the square:

$$(x^2 + 6x \qquad) + y^2 = 0$$
$$(x^2 + 6x + 9) + y^2 = 9$$
$$(x+3)^2 \qquad + y^2 = 3^2$$
$$\implies c(-3,0), \ r = 3.$$

46. We start as before but then divide through by 9 to convert the coefficients of x^2 and y^2 to 1.

$$(9x^2 - 6x \qquad) + (9y^2 + 12y \qquad) = 31$$
$$(x^2 - \tfrac{2}{3}x \qquad) + (y^2 + \tfrac{4}{3}y \qquad) = \tfrac{31}{9}$$
$$(x^2 - \tfrac{2}{3}x + \tfrac{1}{9}) + (y^2 + \tfrac{4}{3}y + \tfrac{4}{9}) = \tfrac{31}{9} + \tfrac{1}{9} + \tfrac{4}{9}$$

$$(x - \tfrac{1}{3})^2 \quad + (y + \tfrac{2}{3})^2 \quad = \frac{36}{9} = 4$$

$$\Rightarrow \quad C(\tfrac{1}{3}, -\tfrac{2}{3}), \text{ and } r = 2.$$

EXERCISES 1.3, page 28

NOTE: In some of the solutions below, m_{AB} will denote the slope of the line through points A and B.

1. $m_{AB} = (18-6)/(-1-(-4)) = 12/(-1+4) = 4.$

4. $m_{AB} = (4-4)/(2+3) = 0.$

7. $m_{AB} = (12-15)/(11-6) = -3/5$ and $m_{CD} = (-5+8)/(-6+1) = -3/5.$ Thus AB $\parallel$ CD.

   Also, $m_{AD} = 20/12 = 5/3$ and $m_{BC} = (-20)/(-12) = 5/3$ and AD $\parallel$ BC. Thus the figure is a parallelogram and since the slopes of adjacent sides are negative reciprocals, these sides are $\perp$ and the figure is a rectangle.

10. Let E, F, G, H be the midpoints of the segments AB, BC, CD, DA, respectively. Then

$$m_{EF} = (\frac{y_2+y_3}{2} - \frac{y_1+y_2}{2}) \Big/ (\frac{x_2+x_3}{2} - \frac{x_1+x_2}{2}) = (y_3-y_1)/(x_3-x_1) \quad (\text{if } x_1 \neq x_3).$$

$$m_{HG} = (\frac{y_3+y_4}{2} - \frac{y_1+y_4}{2}) \Big/ (\frac{x_3+x_4}{2} - \frac{x_1+x_4}{2}) = (y_3-y_1)/(x_3-x_1).$$

   Thus EF $\parallel$ HG since, if $x_1 \neq x_3$, the slopes are equal and if $x_1 = x_3$ both lines are vertical. Similarly $m_{EH} = m_{FG} = (y_2-y_4)/(x_2-x_4)$, if $x_2 \neq x_4$.

13. $m_{AB} = \frac{(-4-(-7))}{(3-(-5))} = \frac{3}{8}.$ Using this slope and the point A(-5,-7) in the point-slope formula (1.16), we obtain: $y-(-7) = (3/8)(x-(-5))$ or $8(y+7) = 3(x+5)$ or $3x - 8y - 41 = 0.$

16. Using m = 6, and the point (-2,0), we get $(y-0) = 6(x-(-2))$ or $y = 6(x+2).$

19. The given line, $2x - 5y = 8$, has slope 2/5 (obtained by writing it as $y = (2/5)x - (8/5)$). Thus the desired slope is -5/2 and the equation is $y + 3 = (-5/2)(x-7).$

22. The desired line must pass through the origin and must have inclination 135°. Since $\tan 135^{\circ} = -1$, the equation is $y = -x.$

25. Rewriting the given line as $y = (3/4)x + 2$, we read off the slope, 3/4, and y-intercept 2.

28. Rewriting as $y = -(8/4)x + 1/4$, the slope is seen to be -2 and y-intercept $\frac{1}{4}$.

31. Rewriting as $y = -(5/4)x + 5$, the slope is seen to be -5/4 and the y-intercept 5.

34. Rewriting as $y = x = (1)x + 0$, the slope is 1 and the y-intercept 0.

37. HINT: Use the 2 points $(a,0)$ and $(0,b)$ to compute the slope; then use (1.16). Dividing $4x - 2y = 6$ by 6 (to get the right side 1) we get

$$\frac{2}{3}x - \frac{1}{3}y = 1 \text{ or } \frac{x}{(3/2)} + \frac{y}{(-3)} = 1.$$

40. $m = \frac{(3t+1)-t}{t-(1-2t)} = \frac{2t+1}{3t-1}$. Because the denominator, $3t-1$, is 0 at $t = 1/3$, we consider the problem separately on the two intervals $t > 1/3$, where $3t-1 > 0$, and $t < 1/3$, where $3t-1 < 0$. First, then, if $t > 1/3$, $m > 4 \iff (2t+1)/(3t-1) > 4 \iff 2t+1 > 12t-4 \iff 5 > 10t$ or $t < 1/2$. Since $t > 1/3$ also, we have $(1/3,1/2)$ as part of the solution set. Secondly, if $t < 1/3$, then $3t-1 < 0$ and $m > 4 \iff 2t+1 < 12t-4 \iff 5 < 10t$ or $t > 1/2$. (The inequality sign changed when we multiplied by $3t-1 < 0$.) Since no numbers satisfy $t < 1/3$ and $t > 1/2$ no new solutions are obtained. Thus $(1/3,1/2)$ is the entire solution set.

EXERCISES 1.4, page 37

1. $f(x) = x^3 + 4x - 3$. (a) $f(1) = 1^3 + 4 \cdot 1 - 3 = 2$. (b) $f(-1) = (-1)^3 + 4(-1) - 3 = -1-4-3 = -8$. (c) $f(0) = 0^3 + 4 \cdot 0 - 3 = -3$. (d) $f(\sqrt{2}) = (\sqrt{2})^3 + 4\sqrt{2} - 3 = 2\sqrt{2} + 4\sqrt{2} - 3 = 6\sqrt{2} - 3$.

4. $f(x) = 1/(x^2+1)$. (a) $f(a) = 1/(a^2+1)$. (b) $f(-a) = 1/((-a)^2+1) = 1/(a^2+1)$. (c) $-f(a) = -1/(a^2+1)$. (d) $f(a+h) = 1/((a+h)^2+1)$. (e) $f(a) + f(h) = (1/(a^2+1)) + (1/(h^2+1))$.

(f) $\frac{f(a+h) - f(a)}{h} = \frac{1}{h}(\frac{1}{(a+h)^2+1} - \frac{1}{a^2+1}) = \frac{1}{h}(\frac{(a^2+1) - ((a+h)^2+1)}{((a+h)^2 + 1)(a^2+1)}) = \frac{1}{h}\frac{a^2+1-a^2-2ah-h^2-1}{((a+h)^2+1)(a^2+1)} = -\frac{(2a+h)}{((a+h)^2+1)(a^2+1)}$.

7. The domain consists of all real numbers x for which $3x-5 \geq 0$. Solving, we get $3x \geq 5$ or $x \geq 5/3$.

10. Here, the domain $= \{x: x^2 - 9 \geq 0\}$. Since $x^2 - 9 \geq 0 \iff x^2 \geq 9 \iff |x| \geq 3$, the domain is $(-\infty,-3] \cup [3,\infty)$.

13. For the 1st part, we solve $f(x) = 4 \iff 7x-5 = 4 \iff 7x = 9 \iff x = 9/7$. For the 2nd part we solve $f(x) = a \iff 7x-5 = a \iff 7x = 5+a \iff x = (5+a)/7$. The range is all of $\mathbb{R}$ for if y is any real number we solve $f(x) = y$ as above to obtain $x = (5+y)/7$. As a check we compute $f((5+y)/7) = 7(\frac{5+y}{7}) - 5 = y$.

16. 1st, $f(x) = 4 \iff 1/x = 4 \iff x = 1/4$. 2nd, $f(x) = a \iff 1/x = a \iff x = 1/a$. The range is $\{y: y \neq 0\}$ for if $y \neq 0$ we solve $f(x) = y$ to obtain $1/x = y$ or $x = 1/y$ and check that $f(1/y) = y$.

19. f is one-to-one since $a \neq b \implies 2a \neq 2b \implies 2a+9 \neq 2b+9$, or $f(a) \neq f(b)$.

22. f is not one-to-one. The graph of f is a parabola opening upward. This sug-
     gests that different x values will yield the same functional value. In par-
     ticular f(0) = f(.5) = -3.

25. f is not one-to-one. If a > 0 then a ≠ -a, but f(a) = a, f(-a) = $|a|$ = -(-a)
     = a.

28. $f(-a) = 7(-a)^4 - (-a)^2 + 7 = 7a^4 - a^2 + 7 = f(a)$... even.

31. f is even since f(a) = f(-a) = 2.

34. f is even since $f(-a) = \sqrt{(-a)^2+1} = \sqrt{a^2+1} = f(a)$.

37. As a polynomial, the domain is all of ℝ. The range is also ℝ since, if
     y ε ℝ, f((3-y)/4) = y.

40. Domain is ℝ; range is {3}; graph is the horizontal line y = 3.

43. Domain = $\{x:4-x^2 \geq 0\}$ = [-2,2]. The graph is that of the equation $y = \sqrt{4-x^2}$,
     i.e., the upper half of the circle $y^2 = 4-x^2$ or $x^2 + y^2 = 4$. Thus the range
     is the set of y-values corresponding to this graph, namely [0,2].

46. Domain = {x:x ≠ 4}; the range = {y:y ≠ 0} for if y ≠ 0 we can solve 1/(4-x) =
     y to obtain x = 4-1/y so that f(4-1/y) = y. The graph is
     obtained as in Example 7, except when x is near 4 the
     ordinate 1/(4-x) is numerically large.

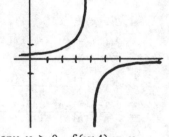

49. Domain = ℝ, range = {y:y ≥ 0} since f(x) ≥ 0 and for any y ≥ 0, f(y+4) = y.

52. Recalling that $|x|$ = x if x ≥ 0, $|x|$ = -x if x < 0,
     f(x) = 2x if x ≥ 0, 0 if x < 0. Thus Domain = ℝ,
     range = [0,∞), and the graph as shown.

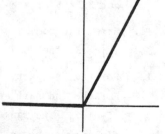

55. From the definition of f, the domain is ℝ; the range is {-1,1} since these
     are the only 2 functional values.

58. From the definition of f, the domain is $\mathbb{R}$, and the range
    is $(0, \infty)$ since if $y > 0$ we can choose $x = -y < 0$ so that
    $f(-y) = y$.

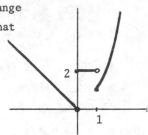

61. For each $x \in (-1,1)$ there correspond 2 points on the circle and thus 2 values
    of y rather than a <u>unique</u> y as required for a function.

EXERCISES 1.5, page 42

1. $(f \underline{+} g)(x) = f(x) \underline{+} g(x) = 3x^2 \underline{+} 1/(2x-3)$. $(fg)(x) = f(x)g(x) = 3x^2(1/(2x-3)) = 3x^2/(2x-3)$. $(f/g)(x) = f(x)/g(x) = 3x^2/(1/(2x-3)) = 3x^2(2x-3)$ (if $x \neq 3/2$).

4. $(f \underline{+} g)(x) = (x^3+3x) \underline{+} (3x^2+1)$. $(fg)(x) = (x^3+3x)(3x^2+1) = 3x^5 + 10x^3 + 3x$. $(f/g)(x) = (x^3+3x)/(3x^2+1)$.

7. If $f(x) = 2x^2 + 5$, $g(x) = 4 - 7x$ then $(f \circ g)(x) = 2g(x)^2 + 5 = 2(4-7x)^2 + 5 = 2(16-56x+49x^2) + 5 = 98x^2 - 112x + 37$; and $(g \circ f)(x) = 4 - 7f(x) = 4 - 7(2x^2+5) = 4 - 14x^2 - 35 = -14x^2 - 31$.

10. If $f(x) = \sqrt{x^2+4}$, $g(x) = 7x^2+1$ then $(f \circ g)(x) = \sqrt{g(x)^2+4} = \sqrt{(7x^2+1)^2 + 4} = \sqrt{49x^4 + 14x^2 + 5}$, and $(g \circ f)(x) = 7f(x)^2 + 1 = 7(x^2+4) + 1 = 7x^2 + 29$.

13. Here, $(f \circ g)(x) = \sqrt{2g(x) + 1} = \sqrt{2x^2+7}$ and $(g \circ f)(x) = f(x)^2 + 3 = (2x+1) + 3 = 2x+4$, for $x \geq -1/2$ (the domain of f).

16. $(f \circ g)(x) = \sqrt[3]{g(x)^2+1} = \sqrt[3]{(x^3 +1)^2+1} = \sqrt[3]{x^6+2x^3+2}$. $(g \circ f)(x) = f(x)^3+1 = (\sqrt[3]{x^2+1})^3+1 = (x^2+1) + 1 = x^2 + 2$.

19. $(f \circ g)(x) = 2g(x) - 3 = 2(\frac{x+3}{2}) - 3 = (x+3) - 3 = x$. $(g \circ f)(x) = \frac{f(x)+3}{2} = \frac{(2x-3)+3}{2} = \frac{2x}{2} = x$.

22. Let $f(x) = a_n x^n + \ldots + a_o$, $g(x) = b_m x^m + \ldots + b_o$, where a_n and $b_m \neq 0$. Then $(fg)(x) = a_n b_m x^{n+m} + \ldots + a_o b_o$ and the degree of the product is $n+m$ since $a_n b_m \neq 0$.

25. HINT: If f is the function, show that $e(x) = (f(x) + f(-x))/2$ is even, that $o(x) = (f(x) - f(-x))/2$ is odd, and that $f(x) = e(x) + o(x)$.

EXERCISES 1.6, page 47

1. $(f \circ g)(x) = 9g(x) + 2 = 9(\frac{1}{9}x - \frac{2}{9}) + 2 = x - 2 + 2 = x$. $(g \circ f)(x) = \frac{f(x)}{9} - \frac{2}{9} = \frac{9x+2}{9} - \frac{2}{9} = x$.

4. With $f(x) = \frac{1}{x-1}$, $g(x) = \frac{1+x}{x} = \frac{1}{x} + 1$, $(f \circ g)(x) = \frac{1}{g(x)-1} = \frac{1}{1/x} = x$, and $(g \circ f)(x) = \frac{1}{f(x)} + 1 = (x-1) + 1 = x$.

7. To find f^{-1} if $f(x) = 6-x^2$, $0 \le x \le \sqrt{6}$, we solve $y = 6-x^2 \iff x^2 = 6-y \implies x = \sqrt{6-y}$, where we chose the positive square root since $x \ge 0$. Thus $f^{-1}(x) = \sqrt{6-x}$, $0 \le x \le 6$.

10. Here, we solve $y = \sqrt{1-4x^2}$, $0 \le x \le 1/2 \implies y^2 = 1-4x^2 \iff 4x^2 = 1-y^2 \implies x = (1/2)\sqrt{1-y^2}$. Thus $f^{-1}(x) = (1/2)\sqrt{1-x^2}$ for $0 \le x \le 1$. (For the original function, the range was $0 \le y \le 1$. This becomes the domain of f^{-1}.)

13. $y = (x^3+8)^5 \iff x^3+8 = y^{1/5} \iff x = (y^{1/5} - 8)^{1/3}$. Thus $f^{-1}(x) = \sqrt[3]{x^{1/5} - 8}$ for all $x \in \mathbb{R}$.

19. Any polynomial function which is not one-to-one won't have an inverse, e.g. $f(x) = x^2$ $(f(1) = f(-1) = 1)$.

EXERCISES 1.7 (Review), page 48

1. $4-3x > 7+2x \iff -3 > 5x \iff -3/5 > x \iff x \in (-\infty, -3/5)$.

4. Observing that $|6x-7| \le 1 \iff -1 \le 6x-7 \le 1 \iff 6 \le 6x \le 8 \iff 1 \le x \le 4/3$, the desired solution set of $|6x-7| > 1$ is the complement of $[1, 4/3]$, namely $(-\infty, 1) \cup (4/3, \infty)$. (This is equivalent to the method of #25, Sec. 1.1.)

7. If $x > 1/3$, both denominators are positive and $1/(3x-1) < 2/(x+5) \iff x+5 < 6x-2 \iff 7 < 5x \iff 7/5 < x$, yielding the partial solution $(7/5, \infty)$. If $-5 < x < 1/3$, $3x-1 < 0$, $x+5 > 0$ and every number in $(-5, 1/3)$ satisfies the inequality. If $x < -5$, both denominators are negative and, as in the first part, the given inequality is equivalent to $7/5 < x$ which is not true for any $x < -5$. Hence $(-5, 1/3) \cup (7/5, \infty)$ is the solution set.

10. $3x - 5y = 10$ is a line with slope $3/5$ and y-intercept -2.

13. $|x+y| = 1$ is satisfied if $x+y = 1$ or $x+y = -1$. Thus the graph consists of these 2 parallel lines.

16. A point (x,y) satisfies $x^2 + y^2 < 1$ if, and only if, its distance from $(0,0)$ is < 1. Thus the graph of W is the set of points inside, but not on, the circle $x^2 + y^2 = 1$.

19. Since $C(-4,-3)$ is 9 units from the vertical line $x = 5$, the radius is 9 and the equation is $(x+4)^2 + (y+3)^2 = 81$.

22. $m_{AC} = (-5-2)/(2-(-4)) = -7/6$ and the equation is $y-6 = (-7/6)(x-3)$ or $7x + 6y - 57 = 0$.

25. Any line parallel to the y-axis has equation x = constant. Since it passes through A(-4,2), its equation is x = -4.

28. Because of the radical in the denominator, the domain is {x: $16 - x^2 > 0$} = (-4,4).

31. If $f(x) = 1/\sqrt{x+1}$, (a) $f(1) = 1/\sqrt{2}$, (b) $f(3) = 1/\sqrt{4} = 1/2$, (c) $f(0) = 1/\sqrt{1} = 1$, (d) $f(\sqrt{2}-1) = 1/\sqrt{(\sqrt{2}-1)+1} = 1/\sqrt{\sqrt{2}} = 1/\sqrt[4]{2}$, (e) $f(-x) = 1/\sqrt{-x+1}$, (f) $-f(x) = -1/\sqrt{x+1}$, (g) $f(x^2) = 1/\sqrt{x^2+1}$, (h) $(f(x))^2 = (1/\sqrt{x+1})^2 = 1/(x+1)$.

34. If x is close to -1, the numbers $-1/(x+1)$ are very large, positive if x < -1, negative if x > -1. If $|x|$ is very large, then $-1/(x+1)$ is nearly 0.

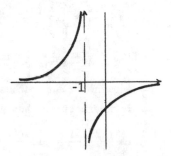

37. $(f+g)(x) = (x^2+4)+\sqrt{2x+5}$, $(fg)(x) = (x^2+4)\sqrt{2x+5}$, $(f/g)(x) = (x^2+4)/\sqrt{2x+5}$, $(f\circ g)(x) = g(x)^2 + 4 = (2x+5) + 4 = 2x+9$ (for $x \geq -5/2$), $(g\circ f)(x) = \sqrt{2f(x) + 5} = \sqrt{2x^2 + 13}$.

40. To show f is one-to-one, let $0 \leq a < b$. Multiplying the inequality first by a and then by b, we obtain $a^2 < ab$, $ab < b^2$ or $a^2 < b^2 \implies 4a^2 + 3 < 4b^2 + 3$ or $f(a) < f(b)$. Thus $f(a) \neq f(b)$, and f is one-to-one. To find f^{-1} we solve $y = 4x^2 + 3$ for x to obtain $x^2 = (1/4)(y-3)$ or $x = (1/2)\sqrt{y-3}$ (positive square root since $x \geq 0$). Thus $f^{-1}(x) = (1/2)\sqrt{x-3}$ for $x \geq 3$.

CHAPTER 2

LIMITS AND CONTINUITY OF FUNCTIONS

EXERCISES 2.1, page 56

1. $\lim\limits_{x\to 2} \dfrac{x^2-4}{x-2} = \lim\limits_{x\to 2} \dfrac{(x+2)(x-2)}{x-2} = \lim\limits_{x\to 2} x+2 = 4.$

4. $\lim\limits_{r\to -3} \dfrac{r^2+2r-3}{r^2+7r+12} = \lim\limits_{r\to -3} \dfrac{(r+3)(r-1)}{(r+3)(r+4)} = \lim\limits_{r\to -3} \dfrac{r-1}{r+4} = -4.$

7. $\lim\limits_{k\to 4} \dfrac{k^2-16}{\sqrt{k}-2} = \lim\limits_{k\to 4} \dfrac{(k+4)(k-4)}{\sqrt{k}-2} = \lim\limits_{k\to 4} \dfrac{(k+4)(\sqrt{k}+2)(\sqrt{k}-2)}{\sqrt{k}-2} = \lim\limits_{k\to 4} (k+4)(\sqrt{k}+2)$

   $= (4+4)(\sqrt{4}+2) = 8 \cdot 4 = 32.$

10. $\lim\limits_{h\to 2} \dfrac{h^3-8}{h^2-4} = \lim\limits_{h\to 2} \dfrac{(h-2)(h^2+2h+4)}{(h-2)(h+2)} = \lim\limits_{h\to 2} \dfrac{h^2+2h+4}{h+2} = \dfrac{4+4+4}{4} = 3.$

13. $\lim\limits_{x\to -3/2} \dfrac{2x+3}{4x^2+12x+9} = \lim\limits_{x\to -3/2} \dfrac{(2x+3)}{(2x+3)^2} = \lim\limits_{x\to -3/2} \dfrac{1}{2x+3}$, which won't exist since the

   denominator approaches 0 as $x \to -\dfrac{3}{2}$ but the numerator remains 1. The values
   of the function thus get arbitrarily large when x is close to -3/2.

16. $\lim\limits_{t\to 1} \dfrac{(1/t)-1}{t-1} = \lim\limits_{t\to 1} \dfrac{(1-t)/t}{(t-1)} = \lim\limits_{t\to 1} \left(-\dfrac{1}{t}\right) = -1.$

19. (a) As in Example 1, with $P(a,f(a)) = P(a,a^3)$ and $Q(x,f(x)) = Q(x,x^3)$, we

   have $m_{PQ} = \dfrac{x^3-a^3}{x-a} = \dfrac{(x-a)(x^2+ax+a^2)}{(x-a)} = x^2 + ax + a^2.$ Then, by (2.2),

   $m = \lim\limits_{x\to a} m_{PQ} = \lim\limits_{x\to a} (x^2 + ax + a^2) = 3a^2.$ (b) With a = 2 we have P(2,8) and

   $m = 3(2^2) = 12$, and the tangent line equation is $y-8 = 12(x-2).$

22. Here a = 4, and $Q(x,\sqrt{x})$, P(4,2). By (2.2), m =

   $\lim\limits_{x\to 4} m_{PQ} = \lim\limits_{x\to 4} \dfrac{\sqrt{x}-2}{x-4} = \lim\limits_{x\to 4} \dfrac{\sqrt{x}-2}{(\sqrt{x}-2)(\sqrt{x}+2)} = \lim\limits_{x\to 4} \dfrac{1}{\sqrt{x}+2} =$

   $\dfrac{1}{\sqrt{4}+2} = \dfrac{1}{2+2} = \dfrac{1}{4}.$ The tangent line equation is y-2 =

   (1/4)(x-4) or x - 4y + 4 = 0.

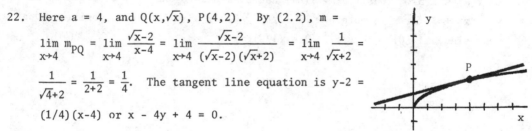

25. If P is the origin, then PQ coincides with y=x if Q is to the right of P
   (x > 0), whereas PQ coincides with y=-x if Q is to the left of P (x < 0).
   As Q approaches P no single limiting line is obtained.

28. HINT: No matter where P and Q are on the graph of f, PQ coincides with the
   graph.

EXERCISES 2.2, page 62

1. Here f(x) = 5x-3, L = 7, a = 2. Then $|f(x) - L| = |(5x-3)-7| = |5x-10| =$

$5|x-2|$. Thus $|f(x) - L| < \varepsilon$ if, and only if, $5|x-2| < \varepsilon$ or $|x-2| < \varepsilon/5$. So if we choose $\delta = \varepsilon/5$, then for numbers x satisfying $0 < |x-2| < \delta$ we have $|f(x) - L| = 5|x-2| < 5\delta = 5(\varepsilon/5) = \varepsilon$, as desired.

4. Here $f(x) = 8x-15$, $L = 17$, $a = 4$. Then $|f(x) - L| = |(8x-15) - 17| = |8x-32| = 8|x-4|$. Thus $|f(x) - L| < \varepsilon$ if and only if $8|x-4| < \varepsilon$ or $|x-4| < \varepsilon/8$. So with $\delta = \varepsilon/8$ and $0 < |x-4| < \delta$ we obtain $|f(x) - L| = 8|x-4| < 8\delta = \varepsilon$.

7. Here $f(x) = 5$, $L = 5$, $a = 3$. $|f(x) - L| = |5-5| = 0$ which is $< \varepsilon$ for all x. Thus δ may be any real number here.

9. $|f(x) - L| = |x-\pi|$.   So with $\delta = \varepsilon$, $0 < |x-\pi| < \delta \Longrightarrow |f(x) - L| < \varepsilon$.

10. As in #7 above, δ may be any real number.

13. If $\lim\limits_{x \to -5} \dfrac{1}{x+5} = L$ existed, then given any $\varepsilon > 0$ we could find $\delta > 0$ such that if $-5-\delta < x < -5+\delta$, $x \neq -5$ (i.e., $-\delta < x+5 < \delta$, $x+5 \neq 0$) then $L - \varepsilon < \dfrac{1}{x+5} < L + \varepsilon$ by (2.7). But this is impossible since $1/(x+5)$ can be made larger than $L + \varepsilon$ by taking x+5 small enough. (For example, if $L > 0$ then $L + \varepsilon > 0$ and $1/(x+5) > L + \varepsilon$ if $0 < x+5 < 1/(L+\varepsilon)$.) Thus, the limit does not exist.

16. Informally, with $f(x) = [x]$ if $a < x_1 < a+1$ then $f(x_1) = a$ whereas if $a-1 < x_2 < a$, then $f(x_2) = a-1$, even if x_1 and x_2 are very close to a. Thus there is no single number which all functional values are close to if x is close to a. (Compare Figure 1.31.) Formally we prove indirectly that no limit exists. If $\lim\limits_{x \to a} f(x) = \lim\limits_{x \to a} [x] = L$ exists where a is an integer, then given any positive $\varepsilon \leq 1/2$ we can find $\delta > 0$ such that if $a-\delta < x < a+\delta$, $x \neq a$ then $|f(x) - L| < \varepsilon \leq 1/2$. Now let x_1 and x_2 satisfy $a-\delta < x_2 < a < x_1 < a+\delta$. Then, as in the 1st sentence $f(x_1) = a$, $f(x_2) = a-1$ and $|f(x_1)-f(x_2)| = |a-(a-1)| = 1$. But, using the triangle inequality we obtain $1 = |f(x_1)-f(x_2)| = |(f(x_1) - L)+(L - f(x_2))| \leq |f(x_1) - L|+|f(x_2) - L| < \varepsilon+\varepsilon < \dfrac{1}{2} + \dfrac{1}{2} = 1$ and we have the absurd result, $1 < 1$, which proves our assumption that the limit existed was wrong.

19. We consider the case where $a > 0$. Let $x_1 = \sqrt{a^2-\varepsilon}$ (so that $f(x_1) = a^2-\varepsilon$) and $x_2 = \sqrt{a^2+\varepsilon}$ (so that $f(x_2) = a^2+\varepsilon$). Then any $x \in (x_1,x_2)$ satisfies the

condition that the point (x,x^2) lies between the
horizontal lines $y = a^2 \pm \varepsilon$. To get an interval of
the proper form let δ be the smaller of x_2-a and
$a-x_1$. (δ can be shown to be x_2-a). Then
$(a-\delta,a+\delta) \subset (x_1,x_2)$ and satisfies the require-
ments.

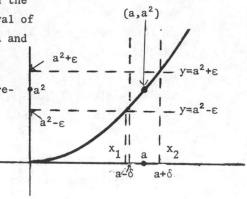

EXERCISES 2.3, page 70

1. $\lim\limits_{x \to -2}$ $(3x^3-2x+7) = 3(-2)^3 - 2(-2) + 7 = -13$ (by 2.18)).

4. $\lim\limits_{t \to -3}$ $(3t+4)(7t-9) = \lim\limits_{t \to -3}$ $(3t+4) \cdot \lim\limits_{t \to -3}$ $(7t-9) = (3(-3)+4)(7(-3)-9) =$

$(-5)(-30) = 150$ (by (2.12(ii)) and (2.9)).

7. $\lim\limits_{x \to 7} 0 = 0$ and 10. $\lim\limits_{x \to 15} \sqrt{2} = \sqrt{2}$ both by (2.8).

13. $\lim\limits_{x \to 2} \dfrac{x-2}{x^3-8} = \lim\limits_{x \to 2} \dfrac{x-2}{(x-2)(x^2+2x+4)} = \lim\limits_{x \to 2} \dfrac{1}{x^2+2x+4} = \dfrac{1}{4+4+4} = \dfrac{1}{12}$.

16. $\lim\limits_{x \to -2} \dfrac{x^3+8}{x^4-16} = \lim\limits_{x \to -2} \dfrac{(x+2)(x^2-2x+4)}{(x+2)(x-2)(x^2+4)} = \lim\limits_{x \to -2} \dfrac{(x^2-2x+4)}{(x-2)(x^2+4)} = \dfrac{12}{(-4)(8)} = -\dfrac{3}{8}$.

19. $\lim\limits_{x \to 1} (\dfrac{x^2}{x-1} - \dfrac{1}{x-1}) = \lim\limits_{x \to 1} \dfrac{x^2-1}{x-1} = \lim\limits_{x \to 1} \dfrac{(x+1)(x-1)}{x-1} = \lim\limits_{x \to 1} x+1 = 2$.

22. $\lim\limits_{x \to -8} \dfrac{16x^{2/3}}{4-x^{4/3}} = \dfrac{16(-8)^{2/3}}{4-(-8)^{4/3}} = \dfrac{16(-2)^2}{4-(-2)^4} = \dfrac{64}{-12} = -\dfrac{16}{3}$.

25. $\lim\limits_{h \to 0} \dfrac{4-\sqrt{16+h}}{h} = \lim\limits_{h \to 0} \dfrac{4-\sqrt{16+h}}{h}\dfrac{(4+\sqrt{16+h})}{(4+\sqrt{16+h})} = \lim\limits_{h \to 0} \dfrac{16-(16+h)}{h(4+\sqrt{16+h})} = \lim\limits_{h \to 0} \dfrac{-1}{4+\sqrt{16+h}} = -\dfrac{1}{8}$.

28. $\lim\limits_{x \to 6} (x+4)^3(x-6)^2 = (6+4)^3(6-6)^2 = (1000)(0) = 0$.

31. $\lim\limits_{t \to -1} \dfrac{(4t^2+5t-3)^3}{(6t+5)^4} = \dfrac{(4(-1)^2+5(-1)-3)^3}{(6(-1)+5)^4} = \dfrac{(4-5-3)^3}{(-1)^4} = (-4)^3 = -64$.

34. First, from the identity $a^3-b^3 = (a-b)(a^2+ba+b^2)$ we obtain, with $a = \sqrt[3]{x}$ and $b = 2$, $x-8 = (\sqrt[3]{x}-2)(\sqrt[3]{x^2}+2\sqrt[3]{x}+4)$. Thus $\lim\limits_{x \to 8} \dfrac{x-8}{\sqrt[3]{x}-2} = \lim\limits_{x \to 8} (\sqrt[3]{x^2}+2\sqrt[3]{x}+4) = \sqrt[3]{64} + 2\sqrt[3]{8} + 4 = 4 + 2(2) + 4 = 12$.

37. HINT: With $r = m/n$ in lowest terms, recall that $x^r = x^{m/n} = (\sqrt[n]{x})^m$. Then use (2.14) and (2.20) and the conditions there.

40. HINT: $x \neq 0 \implies x^4 + 4x^2 > 0 \implies \sqrt{x^4+4x^2+7} > \sqrt{7} > 1 \implies |x|/\sqrt{x^4+4x^2+7} < |x|$.

43. If $\lim\limits_{x \to a} f(x) = L < 0$, then, if $F(x) = -f(x)$, $\lim\limits_{x \to a} F(x) = -L$ by (2.15) with

   $c = -1$. By (2.10) there is an open interval I containing a on which $F(x) > 0$
   (except possibly at $x = a$). Thus $F(x) = -f(x) > 0$ or $f(x) < 0$ there and the
   proof is complete.

EXERCISES 2.4, page 75

1. $\lim\limits_{x \to 0^+} (4 + \sqrt{x}) = 4 + \sqrt{0} = 4$.

4. Note that as $x \to 5/2^-$, $x < 5/2$, $5 - 2x > 0$ and such x are in the domain of f.
   $\lim\limits_{x \to 5/2^-} (\sqrt{5-2x} - x^2) = \sqrt{5-2(5/2)} - (5/2)^2 = 0 - 25/4$. (The right-hand limit,

   however, does not exist for if $x > 5/2$ then $5-2x < 0$, and f is not defined
   for such x.)

7. $\lim\limits_{x \to 3^-} \dfrac{|x-3|}{x-3} = \lim\limits_{x \to 3^-} \dfrac{-(x-3)}{x-3} = -1$ since as $x \to 3^-$, $x < 3$, $x-3 < 0$ and, therefore,

   $|x-3| = -(x-3)$.

10. $\lim\limits_{x \to -10^-} \dfrac{x+10}{\sqrt{(x+10)^2}} = \lim\limits_{x \to -10^-} \dfrac{x+10}{|x+10|} = \lim\limits_{x \to -10^-} \dfrac{x+10}{-(x+10)} = -1$ since, if $x < -10$,

   $x+10$ is negative and $|x+10| = -(x+10)$.

13. $\lim\limits_{x \to -7^+} \dfrac{x+7}{|x+7|} = \lim\limits_{x \to -7^+} \dfrac{x+7}{x+7} = \lim\limits_{x \to -7^+} 1 = 1$ since if $x > -7$ then $x+7 > 0$ and $|x+7| =$

   $x+7$.

16. The limit does not exist. If $x < 8$ and $x-8$ is small then $1/(x-8)$ is very
    large and negative. See also Exercise 38, Section 2.3 of the text with
    $f(x) = 1$, $g(x) = x-8$ and one-sided limits.

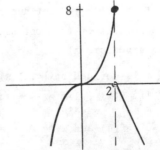

18. $\lim\limits_{x \to 2^-} f(x) = \lim\limits_{x \to 2^-} x^3 = 8$.

    $\lim\limits_{x \to 2^+} f(x) = \lim\limits_{x \to 2^+} (4-2x) = 0$.

19. $\lim\limits_{x \to -3^+} f(x) = \lim\limits_{x \to -3^+} \sqrt[3]{x+2} = \sqrt[3]{-3+2} = -1$ since if $x > -3$ the second formula in

    the definition of f must be used. $\lim\limits_{x \to -3^-} f(x) = \lim\limits_{x \to -3^-} \dfrac{1}{2-3x} = \dfrac{1}{2-3(-3)} = \dfrac{1}{11}$.

    Since the left and right limits differ, $\lim\limits_{x \to -3} f(x)$ does not exist.

22. For n < x < n+1, f(x) = 1 and, thus, $\lim\limits_{x \to n^+} f(x) = \lim\limits_{x \to n^+} 1 = 1$. Similarly, if

n-1 < x < n, f(x) = 1 and $\lim\limits_{x \to n^-} f(x) = \lim\limits_{x \to n^-} 1 = 1$. (The value, f(n) = 0, has

__absolutely__ __nothing__ to do with these limits since x ≠ n is required for both
limits.)

25. Recall that if n < x < n+1, then [x] = n so that $\lim\limits_{x \to n^+} [x] = n$.   If n-1 < x

< n, then [x] = n-1 and $\lim\limits_{x \to n^-} [x] = n-1$. Thus $\lim\limits_{x \to n^-} x-[x] = n-(n-1) = 1$, and

$\lim\limits_{x \to n^+} x-[x] = n-n = 0$.

29. Sketch for right-hand limit.   Sketch for left-hand limit.

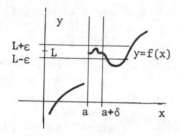

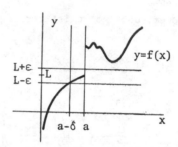

EXERCISES 2.5, page 83

1. $\lim\limits_{x \to 4} (\sqrt{2x-5} + 3x) = \lim\limits_{x \to 4} \sqrt{2x-5} + \lim\limits_{x \to 4} 3x = \sqrt{3} + 12 = f(4)$.

4. f(x) = 1/x is continuous at every number a ≠ 0 by (2.30) since it is the
    quotient of the continuous functions 1 and x, and if a ≠ 0, the denominator
    is nonzero at a.

7. f is defined on [4,8] since x-4 ≥ 0 for all x there. If c ε (4,8) then
    $\lim\limits_{x \to c} f(x) = \lim\limits_{x \to c} \sqrt{x-4} = \sqrt{c-4} = f(c)$ by (2.9) and (2.22). Similarly $\lim\limits_{x \to 4^+} \sqrt{x-4} =$
    0 = f(4) and $\lim\limits_{x \to 8^-} \sqrt{x-4} = 2 = f(8)$. Thus f is continuous on [4,8].

10. If c ε (1,3) then $\lim\limits_{x \to c} f(x) = \lim\limits_{x \to c} \dfrac{1}{x-1} = \dfrac{1}{c-1} = f(c)$ by (2.9) and (2.12)(iii),
     valid since 1 < c < 3 ⟹ 0 < c-1.

13. The domain of f is {x: 2x-3 ≥ 0} = [3/2,∞). If a ε (3/2,∞), $\lim\limits_{x \to a} f(x) =$

$\sqrt{2a-3} + a^2 = f(a)$. Also, $\lim\limits_{x \to 3/2^+} f(x) = \sqrt{2(3/2)-3} + (3/2)^2 = 9/4 = f(3/2)$.

Thus f is continuous throughout its domain.

16. The domain of f is (-1,1), and f is continuous there by the quotient theorem for continuous functions, valid since the numerator is a polynomial and the denominator is continuous by (2.22).

19. As a rational function, f is continuous at every number in its domain. Since the denominator $x^3 - x^2 = x^2(x-1)$ is 0 when x=0 and x=1, the domain of f is $\{x: x \neq 0, 1\} = (-\infty, 0) \cup (0, 1) \cup (1, \infty)$.

22. The numerator, $\sqrt{9-x}$, is continuous by (2.22) throughout its domain which is $\{x: 9-x \geq 0\} = \{x: 9 \geq x\} = (-\infty, 9]$. Similarly the denominator, $\sqrt{x-6}$, is continuous on $[6, \infty)$ but is 0 at x = 6. Thus both are continuous and the denominator is nonzero on $(-\infty, 9] \cap (6, \infty) = (6, 9]$ where f is continuous by the quotient theorem.

25. For the function of #23, Section 4, f is discontinuous at every nonzero integer n since $\lim\limits_{x \to n} f(x) = \lim\limits_{x \to n} 0 = 0 \neq n = f(n)$. (f(x) = 0 if n-1 < x < n+1, $x \neq n$.)

28. For the function of #26, Section 4, we have $\lim\limits_{x \to 3} f(x) = \lim\limits_{x \to 3} \dfrac{x^2-9}{x-3} =$
$\lim\limits_{x \to 3} \dfrac{(x+3)(x-3)}{x-3} = \lim\limits_{x \to 3} x+3 = 6 \neq 5 = f(3)$. Thus f is discontinuous at x = 3. f is continuous at all $x \neq 3$ by the quotient theorem.

31. No. The left-hand limit, as $x \to 3^-$, is -1 by #7, Section 2.4. However, $\lim\limits_{x \to 3^+} \dfrac{|x-3|}{x-3} = \lim\limits_{x \to 3^+} \dfrac{x-3}{x-3} = 1$ since as $x \to 3^+$, x > 3 and x-3 > 0. Since the left and right-hand limits differ, $\lim\limits_{x \to 3} f(x)$ does not exist.

37. Let w be given between f(0) = 4 and f(1) = 9. We must show there is a number c between 0 and 1 such that f(c) = w. Substituting, we get $c^2 + 4c + 4 = w$ $\Longleftrightarrow c^2 + 4c + (4-w) = 0$, and by the quadratic formula,
$c = \dfrac{-4 \pm \sqrt{4^2 - 4(4-w)}}{2} = \dfrac{-4 \pm \sqrt{4w}}{2} = -2 \pm \sqrt{w}$.
We must choose the + sign, or else c < 0. Thus $c = -2 + \sqrt{w}$. Note that $4 < w < 9 \Longrightarrow 2 < \sqrt{w} < 3 \Longrightarrow 0 < -2 + \sqrt{w} = c < 1$, as required.

40. Let $f(x) = x^5 - 3x^4 - 2x^3 - x + 1$. Then f(0) = 1, f(1) = -4 and f is continuous on [0,1]. If we choose w = 0, w lies between f(0) and f(1), and, by the intermediate value theorem, there exists a number $c \in (0,1)$ such that f(c) = 0. This number, c, is a solution of the given equation.

EXERCISES 2.6 (Review), page 85

1. $\lim\limits_{x \to 3} \dfrac{5x+11}{\sqrt{x+1}} = \dfrac{5(3)+11}{\sqrt{3+1}} = \dfrac{26}{2} = 13.$

4. $\lim\limits_{x \to 4^-} (x - \sqrt{16-x^2}) = 4 - \sqrt{16-4^2} = 4.$ (The right-hand limit, however, does not

    exist.)

7. $\lim\limits_{x \to 2} \dfrac{x^4-16}{x^2-x-2} = \lim\limits_{x \to 2} \dfrac{(x+2)(x-2)(x^2+4)}{(x-2)(x+1)} = \lim\limits_{x \to 2} \dfrac{(x+2)(x^2+4)}{x+1} = \dfrac{(4)(8)}{3} = \dfrac{32}{3}.$

10. $\lim\limits_{x \to 5} \dfrac{(1/x)-(1/5)}{x-5} = \lim\limits_{x \to 5} \dfrac{(5-x)/5x}{x-5} = \lim\limits_{x \to 5} \dfrac{-1}{5x} = -\dfrac{1}{25}.$

13. $\lim\limits_{x \to 3^+} \dfrac{3-x}{|3-x|} = \lim\limits_{x \to 3^+} \dfrac{3-x}{-(3-x)} = -1$ since $x > 3 \implies 3-x < 0 \implies |3-x| = -(3-x).$

16. $\lim\limits_{x \to -3} \sqrt[3]{\dfrac{x+3}{x^3+27}} = \lim\limits_{x \to -3} \sqrt[3]{\dfrac{x+3}{(x+3)(x^2-3x+9)}} = \lim\limits_{x \to -3} \sqrt[3]{\dfrac{1}{x^2-3x+9}} = \sqrt[3]{\dfrac{1}{9+9+9}} = \sqrt[3]{\dfrac{1}{27}} = \dfrac{1}{3}.$

19. $\lim\limits_{x \to 2^+} \dfrac{|x-2|}{(2-x)} = \lim\limits_{x \to 2^+} \dfrac{(x-2)}{(2-x)} = -1$ $(x > 2 \implies x-2 > 0 \implies |x-2| = (x-2)).$

22. If $2 < x < 3$, then $[x] = 2$. Thus $\lim\limits_{x \to 3^-} [x] = 2$ and $\lim\limits_{x \to 3^-} [x] - x^2 = 2-9 = -7.$

    (However, in #21, $\lim\limits_{x \to 3^+} [x] = 3$ and $\lim\limits_{x \to 3^+} [x] - x^2 = 3-9 = -6.$)

25. If c is any real number, $\lim\limits_{x \to c} f(x) = \lim\limits_{x \to c} (2x^4 - \sqrt[3]{x} + 1) = 2c^4 - \sqrt[3]{c} + 1 = f(c)$

    by limit theorems of Section 3. Thus f is continuous.

28. $\sqrt{x}$ requires $x \geq 0$, and x^2-1 in the denominator means $x = 1$ must be excluded.

    Thus the domain of f is $[0,1) \cup (1,\infty)$ where f is continuous by the theorems

    of Sections 3 and 5.

31. f is a rational function which is continuous everywhere except at the points

    where the denominator is 0. Here $x^2-2x = x(x-2) = 0$ at $x = 0$ and $x = 2$ where

    f is discontinuous.

34. Let $f(x) = x^5 + 7x^2 - 3x - 5$. Then $f(-2) = -32 + 28 + 6 - 5 = -3$, $f(-1) =$

    $-1 + 7 + 3 - 5 = 4$, and f is continuous on $[-2,-1]$. Choosing $w = 0$ and

    noting that $f(-2) < 0 < f(-1)$, by the intermediate value theorem there exists

    a number c between -2 and -1 such that $f(c) = 0$. Thus, c is the desired

    solution of the given equation.

CHAPTER 3

THE DERIVATIVE

EXERCISES 3.1, page 92

1. By (3.1) the slope at $P(a,f(a)) = P(a,2-a^3)$ is

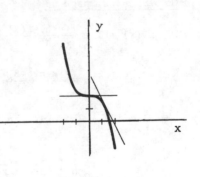

$$m = \lim_{h\to 0} \frac{f(a+h)-f(a)}{h} = \lim_{h\to 0} \frac{(2-(a+h)^3)-(2-a^3)}{h}$$

$$= \lim_{h\to 0} \frac{-a^3 - 3a^2h - 3ah^2 - h^3 + a^3}{h}$$

$$= \lim_{h\to 0} \frac{-h(3a^2 + 3ah + h^2)}{h} = \lim_{h\to 0} -(3a^2 + 3ah + h^2)$$

$$= -3a^2.$$

4. The slope at $P(a,f(a)) = P(a,1/a-1)$ is

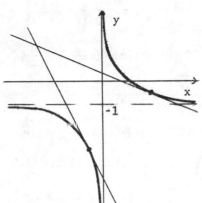

$$\lim_{h\to 0} \frac{(1/(a+h) - 1)-(1/a) - 1)}{h}$$

$$= \lim_{h\to 0} \frac{1}{h}\left(\frac{1}{a+h} - \frac{1}{a}\right) = \lim_{h\to 0} \frac{1}{h} \cdot \frac{a-(a+h)}{(a+h)a}$$

$$= \lim_{h\to 0} \frac{-h}{h(a+h)a} = \lim_{h\to 0} \frac{-1}{(a+h)a} = -\frac{1}{a^2} .$$

7. The velocity at $t = a$ is $\lim_{h\to 0} \dfrac{f(a+h)-f(a)}{h}$ where $f(t)$ is the distance function.

If $v(a)$ is the velocity at $t = a$, we have, with $f(t) = 112t - 16t^2$,

$$v(a) = \lim_{h\to 0} \frac{[112(a+h) - 16(a+h)^2] - [112a - 16a^2]}{h}$$

$$= \lim_{h\to 0} \frac{112h - 32ah - 16h^2}{h} = \lim_{h\to 0} 112 - 32a - 16h$$

$$= 112 - 32a \text{ ft/sec.}$$

Thus $v(2) = 112-64 = 48$, $v(3) = 112-96 = 16$, $v(4) = 112-128 = -16$ ft/sec.
(i.e. downward motion at $t = 4$). The maximum height occurs when $v(a) = 0 \iff$
$112-32a = 0 \iff a = 112/32 = 7/2$ sec. The object hits the ground when the
distance above it is 0, i.e. when $f(t) = 0$. Solving: $112t - 16t^2 = 0 \iff$
$16t(7-t) = 0 \iff t = 0$ or $t = 7$. So, it leaves the ground at $t=0$ and re-
turns to strike it at $t=7$. Impact velocity is $v(7)=112-32(7) = -112$ ft/sec.

10. If $f(t) = c$, then the velocity at any time $t = a$ is $\lim_{h\to 0} \dfrac{f(a+h) - f(a)}{h} =$

$\lim_{h\to 0} \dfrac{c-c}{h} = \lim_{h\to 0} 0 = 0$. The "motion" of the particle is to remain motionless at

the point corresponding to the number c on the number line.

EXERCISES 3.2, page 98

1. $f'(x) = \lim_{h\to 0} \frac{f(x+h)-f(x)}{h} = \lim_{h\to 0} \frac{37-37}{h} = \lim_{h\to 0} 0 = 0$ for all x.

4. $f'(x) = \lim_{h\to 0} \frac{(7(x+h)^2-5)-(7x^2-5)}{h} = \lim_{h\to 0} \frac{7(x^2+2xh+h^2)-7x^2}{h} = \lim_{h\to 0} \frac{14xh+7h^2}{h}$

   $= \lim_{h\to 0} 14x + 7h = 14x$ for all x.

7. $f'(x) = \lim_{h\to 0} \frac{1}{h} [\frac{1}{x+h-2} - \frac{1}{x-2}] = \lim_{h\to 0} \frac{1}{h} [\frac{(x-2)-(x+h-2)}{(x+h-2)(x-2)}] = \lim_{h\to 0} \frac{1}{h} \frac{-h}{(x+h-2)(x-2)}$

   $= \lim_{h\to 0} \frac{-1}{(x+h-2)(x-2)} = \frac{1}{(x-2)^2}$, for $x \neq 2$.

10. $f'(x) = \lim_{h\to 0} \frac{1}{h} [\frac{1}{2(x+h)} - \frac{1}{2x}] = \frac{1}{2} \lim_{h\to 0} \frac{1}{h} [\frac{x-(x+h)}{(x+h)x}] = \frac{1}{2} \lim_{h\to 0} \frac{1}{h} \frac{-h}{(x+h)x}$

   $= \frac{1}{2} \lim_{h\to 0} \frac{-1}{(x+h)x} = - \frac{1}{2x^2}$, for $x \neq 0$.

13. If $y = f(x)$, then $D_x y$ is just $f'(x)$. Thus,

   $D_x y = \lim_{h\to 0} \frac{[2(x+h)^3 - 4(x+h) + 1] - [2x^3 - 4x + 1]}{h}$

   $= \lim_{h\to 0} \frac{[(2x^3 + 6x^2h + 6xh^2 + 2h^3) - (4x + 4h) + 1] - [2x^3 - 4x + 1]}{h}$

   $= \lim_{h\to 0} \frac{6x^2h + 6xh^2 + 2h^3 - 4h}{h} = \lim_{h\to 0} 6x^2 + 6xh + 2h^2 - 4 = 6x^2 - 4.$

16. $f'(a) = \lim_{x\to a} \frac{f(x)-f(a)}{x-a} = \lim_{x\to a} \frac{\sqrt{2}x - \sqrt{2}a}{x-a} = \lim_{x\to a} \frac{\sqrt{2}(x-a)}{x-a} = \lim_{x\to a} \sqrt{2} = \sqrt{2}.$

19. $f'(a) = \lim_{x\to a} \frac{[1/(x+5) - 1/(a+5)]}{x-a} = \lim_{x\to a} \frac{(a+5) - (x+5)}{(x-a)(x+5)(a+5)} = \lim_{x\to a} \frac{a-x}{(x-a)(x+5)(a+5)}$

   $= \lim_{x\to a} \frac{-1}{(x+5)(a+5)} = - \frac{1}{(a+5)^2}$.

22. First note that $\lim_{h\to 0^+} [5+h] = 5$ and $\lim_{h\to 0^-} [5+h] = 4$. Then, $\lim_{h\to 0^+} \frac{f(5+h)-f(5)}{h}$

   $= \lim_{h\to 0^+} \frac{[5+h]-[5]}{h} = \lim_{h\to 0} \frac{5-5}{h} = 0$, but $\lim_{h\to 0^-} \frac{[5+h]-[5]}{h} = \lim_{h\to 0^-} \frac{4-5}{h} = \lim_{h\to 0^-} \frac{-1}{5}$

   which does not exist. Both limits must exist and be equal in order for f to be differentiable at 5. Thus f is not differentiable there.

EXERCISES 3.3, page 106

1. $f'(x) = D_x(10x^2 + 9x - 4) = D_x(10x^2) + D_x(9x) - D_x(4)$

$= 10 \, D_x(x^2) + 9 \, D_x(x) - D_x(4) = 10(2x) + 9(1) - 0 = 20x + 9.$

4. $f'(t) = D_t(12 - 3t^4 + 4t^6) = D_t(12) - D_t(3t^4) + D_t(4t^6) = D_t(12) - 3 \, D_t(t^4)$

     $+ \, 4 \, D_t(t^6) = 0 - 3(4t^3) + 4(6t^5) = -12t^3 + 24t^5.$

7. $h'(r) = D_r(r^2(3r^4 - 7r + 2)) = D_r(3r^6 - 7r^3 + 2r^2) = 3(6r^5) - 7(3r^2) + 2(2r)$

     $= 18r^5 - 21r^2 + 4r.$

10. $h'(x) = D_x(\dfrac{8x^2 - 6x + 11}{x-1}) = \dfrac{(x-1)D_x(8x^2 - 6x + 11) - (8x^2 - 6x + 11)D_x(x-1)}{(x-1)^2}$

     $= [(x-1)(16x-6) - (8x^2 - 6x + 11)(1)] / (x-1)^2$

     $= (16x^2 - 6x - 16x + 6 - 8x^2 + 6x - 11)/(x-1)^2$

     $= (8x^2 - 16x - 5)/(x-1)^2$

13. $D_x(3x^3 - 2x^2 + 4x - 7) = D_x(3x^3) + D_x(-2x^2) + D_x(4x) + D_x(-7) = 3(3x^2) +$

     $(-2)(2x) + 4(1) + 0 = 9x^2 - 4x + 4.$

16. $D_x(2x + 1/2x) = D_x(2x) + D_x((1/2)x^{-1}) = 2 + (1/2)(-1)x^{-2} = 2 - \dfrac{1}{2x^2} \, .$

19. $D_v\dfrac{(v^3-1)}{(v^3+1)} = \dfrac{(v^3+1)D_v(v^3-1) - (v^3-1)D_v(v^3+1)}{(v^3+1)^2} = \dfrac{(v^3+1)(3v^2) - (v^3-1)(3v^2)}{(v^3+1)^2}$

     $= \dfrac{6v^2}{(v^3+1)^2} \, .$

22. $D_x(1 + 1/x + 1/x^2 + 1/x^3) = D_x(1) + D_x(x^{-1}) + D_x(x^{-2}) + D_x(x^{-3})$

     $= 0 - x^{-2} - 2x^{-3} - 3x^{-4} = -(1/x^2 + 2/x^3 + 3/x^4).$

25. $D_s((3x)^{-4}) = D_s(3^{-4}s^{-4}) = 3^{-4}D_s(s^{-4}) = 3^{-4}(-4)s^{-5} = -4/(81s^5) = -(4/81)s^{-5}.$

28.   First note that if $f = g$ in the product rule we get $(f^2)' = 2ff'$. Here

     $f(r) = (5r-4)^{-1} = \dfrac{1}{5r-4}$ and $f'(r) = \dfrac{(5r-4)D_r(1) - 1 \, D_r(5r-4)}{(5r-4)^2} = \dfrac{-5}{(5r-4)^2} \, .$

     Combining these calculations gives us $D_r(5r-4)^{-2} = 2(5r-4)^{-1}(-5(5r-4)^{-2})$

     $= -10(5r-4)^{-3}.$

31. $D_x(\dfrac{2x^3 - 7x^2 + 4x + 3}{x^2}) = D_x(2x - 7 + 4x^{-1} + 3x^{-2}) = 2 - 4x^{-2} - 6x^{-3}.$

34.   (a) $D_x(\dfrac{x^2+1}{x^4}) = \dfrac{x^4(2x) - (x^2+1)(4x^3)}{x^8} = -\dfrac{2x^2 + 4}{x^5} \, .$

     (b) $D_x((x^2+1)x^{-4}) = (x^2+1)(-4x^{-5}) + x^{-4}(2x) = -2x^{-3} - 4x^{-5} = -(2x^2 + 4)x^{-5},$

       the same as (a).

37.   Grouping fgh as (fg)h we have by the product rule for 2 functions: $D_x(fgh)$

     $= D_x((fg)h) = (fg)h' + h(fg)' = fgh' + h(fg' + g'f) = fgh' + fhg' + hgf'.$

40. Using #37 with $f(x) = 3x^4 - 10x^2 + 8$, $g(x) = 2x^2 - 10$, $h(x) = 6x + 7$, we obtain $y' = (3x^4 - 10x^2 + 8)(2x^2 - 10)(6) + (3x^4 - 10x^2 + 8)(6x + 7)(4x) + (6x + 7)(2x^2 - 10)(12x^3 - 20x)$.

43. Here $y' = 3x^2 + 4x - 4$.   (a) The tangent line is horizontal when $y' = 0$. $3x^2 + 4x - 4 = (3x-2)(x+2) = 0$ if $x = 2/3$ and $x = -2$.   (b) The given line, $2y + 8x - 5 = 0$ has slope -4. The tangent line is parallel to it when $y' = -4$. $3x^2 + 4x - 4 = -4 \iff 3x^2 + 4x = x(3x+4) = 0 \implies x = 0$ and $x = -4/3$.

46. The velocity $v(t) = D_t f(t) = D_t(3t^5 - 5t^3) = 15t^4 - 15t^2 = 15t^2(t^2-1)$.

    (a) Motion is in the positive direction when $v(t) > 0$. Since $15t^2 \geq 0$, this reduces to the condition $t^2 - 1 > 0$ or $t^2 > 1$, which has solution $t > 1$ or $t < -1$.

    (b) Motion is in the negative direction when $v(t) < 0$ or $-1 < t < 1$.

    (c) $v(t) = 0$ at $t = -1, 0, 1$.

49. $f(x) = x^3 - x^2 + x + 1 \implies f'(x) = 3x^2 - 2x + 1$. To find where the graphs of $f(x)$ and $f'(x)$ cross (intersect), we set $f(x) = f'(x)$ and solve for x. $f(x) = f'(x) \iff x^3 - x^2 + x + 1 = 3x^2 - 2x + 1 \iff x^3 - 4x^2 + 3x = 0 \iff x(x^2 - 4x + 3) = x(x-1)(x-3) = 0$.   The solutions are $x = 0, 1, 3$.

52. Here, $y' = 3x^2$. The given line has slope $16/3$. A tangent line to $y = x^3$ is parallel to the given line when $y' = 16/3 \iff 3x^2 = 16/3 \iff x^2 = 16/9 \iff x = \pm 4/3$. With $x = 4/3$, $y = x^3 = 64/27$, and, since the slope is $16/3$, the tangent line here is: $y - \dfrac{64}{27} = \dfrac{16}{3}(x - \dfrac{4}{3})$, or $144x - 27y = 128$. With $x = -4/3$, $y = -64/27$ and the other tangent line is $y + \dfrac{64}{27} = \dfrac{16}{3}(x + \dfrac{4}{3})$ or $144x - 27y = -128$.

EXERCISES 3.4, page 113

1. $\Delta y = f(x+\Delta x) - f(x) = f(2-.2) - f(2) = f(1.8) - f(2) = (2(1.8)^2 - 4(1.8) + 5) - (2(2)^2 - 4(2) + 5) = 6.48 - 7.20 + 5 - 8 + 8 - 5 = -0.72$.

4. $\Delta y = f(0-.03) - f(0) = \dfrac{-.03 + 2}{-.03 + 3} - \dfrac{2}{3} = \dfrac{1.97}{2.97} - \dfrac{2}{3} = \dfrac{5.91 - 5.94}{(2.97)3} = \dfrac{-.03}{3(2.97)}$

   $= -\dfrac{1}{297} \approx -.0034$.

7. (a) $\Delta y = \dfrac{1}{x + \Delta x} - \dfrac{1}{x} = \dfrac{x - (x + \Delta x)}{x(x + \Delta x)} = -\dfrac{\Delta x}{x(x + \Delta x)}$.

   (b) $dy = D_x y \, \Delta x = -\dfrac{1}{x^2} \Delta x$.

   (c) $dy - \Delta y = -\dfrac{\Delta x}{x^2} + \dfrac{\Delta x}{x(x+\Delta x)} = \dfrac{\Delta x}{x}(\dfrac{1}{x+\Delta x} - \dfrac{1}{x}) = \dfrac{\Delta x}{x}(\dfrac{-\Delta x}{x(x+\Delta x)}) = -\dfrac{\Delta x^2}{x^2(x+\Delta x)}$.

10. (a) $\Delta y = f(x+\Delta x) - f(x) = 8 - 8 = 0$. (b) $D_x y = D_x 8 = 0 \Longrightarrow dy = D_x y \, \Delta x = 0(\Delta x) = 0$. (c) $\Delta y - dy = 0 - 0 = 0$.

13. Let $y = f(x)$. The change $\Delta y = f(1.03) - f(1)$ is to be estimated. Here $x = 1$, $x+\Delta x = 1.03$, so $\Delta x = .03$. By (3.33), $\Delta y \approx dy = f'(x)\Delta x = f'(1)(.03)$. $f'(x) = 20x^4 - 24x^3 + 6x \Longrightarrow f'(1) = 2$. Thus $\Delta y \approx 2(.03) = .06$.

16. $dS = F'(t)\Delta t = D_t\left(\dfrac{1}{2-t^2}\right)\Delta t = \dfrac{2t}{(2-t^2)^2}\,\Delta t$. If t changes from 1 to 1.02, we take $t = 1$, $\Delta t = .02$. Then $\Delta S \approx dS = F'(1)\Delta t = \dfrac{2}{(2-1^2)^2}(.02) = 2(.02) = .04$.

18. If A is the area of the square of side x, then $A = x^2$ where $x = 1'$ with error Δx where $|\Delta x| \leq \dfrac{1}{16}'' = \dfrac{1}{192}'$. The actual error $\Delta A \approx dA = 2x\Delta x$ so that $|\Delta A| \approx |dA| = 2|\Delta x| \leq \dfrac{2}{192} = \dfrac{1}{96}$ sq. ft. The last figure is the maximum possible error. The average error is $\approx dA/A \approx (1/96)/1^2 = 1/96$. The % error is $\approx \dfrac{1}{96} \cdot 100 \approx 1.04\%$.

19. If V is the volume of a cube of edge x, then $V = x^3$. Here $x = 10$, $\Delta x = .1$. Thus $\Delta V \approx dV = 3x^2\Delta x = 300(.1) = 30$ in^3. Exactly, $\Delta V = 10.1^3 - 10^3 = 1030.301 - 1000 = 30.301$ in^3.

22. If r is the radius of the cylinder and hemisphere, the silo volume is the volume of the cylinder + the volume of the hemisphere = $\pi r^2 h + \dfrac{2}{3}\pi r^3$. If C is the circumference of the cylinder then $C = 2\pi r$, or $r = C/2\pi$, and $C = 30 \pm .5$ ft. So. in terms of C, using $h = 50$, the silo volume is $V(C) = 50\pi\left(\dfrac{C}{2\pi}\right)^2 + \dfrac{2}{3}\pi\left(\dfrac{C}{2\pi}\right)^3 = \dfrac{25}{2\pi}C^2 +$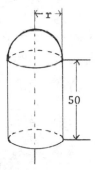

$\dfrac{1}{12\pi^2}C^3$ and $V(30) = \dfrac{25}{2\pi}(30)^2 + \dfrac{1}{12\pi^2}(30)^3 = \dfrac{900}{2\pi^2}(25\pi + 5)$ ≈ 3808.96 ft^3. The error in this calculation is $\Delta V \approx dV = V'(C)\Delta C$. To get the maximum error we can proceed as in #18 above, or, alternately, take $C = 30$, the measured value, and $\Delta C = 0.5$, the maximum error in C. $V'(C) = \dfrac{25C}{\pi}$ $+ \dfrac{C^2}{4\pi^2} \Longrightarrow$ the maximum error in V is $V'(30)(.5) = \left[\dfrac{25(30)}{\pi} + \dfrac{900}{4\pi^2}\right](.5) = \dfrac{150}{4\pi^2}(10\pi + 3) \approx 130.8$. The average error is $\approx \dfrac{dV}{V} = \dfrac{(10\pi + 3)}{60(5\pi + 1)} \approx 0.0343$, and the % error is $\approx 3.43\%$.

25. Here, we are given the % change in F and are to find the Δs that produces it. Recall: % change $= \dfrac{\Delta F}{F}(100) \approx \dfrac{dF}{F}(100) = \dfrac{F'(s)\Delta s}{F}(100)$. Let $K = gm_1 m_2$, a constant. Then $F(s) = Ks^{-2}$, and $F'(s) = -2Ks^{-3}$, so that $\dfrac{F'(s)}{F} = \dfrac{-2Ks^{-3}}{Ks^{-2}} = -\dfrac{2}{s}$. Using this above, setting $s = 20$ and setting the result equal to 10 (the de-

aired % change) we obtain: $\frac{-2\Delta s}{20}(100) = 10 \iff -10\Delta s = 10 \iff \Delta s = -1.$

28. Let $f(x) = x^4 - 3x^3 + 4x^2 - 5$ so that $f(2.01)$ is desired. Taking $x = 2$, $\Delta x = .01$, we know $\Delta y = f(2.01) - f(2) \approx dy = f'(2)(.01)$ or $N = f(2.01) \approx f'(2)(.01) + f(2)$. Now, $f'(x) = 4x^3 - 9x^2 + 8x$, $f'(2) = 4(8) - 9(4) + 8(2) = 12$ and $f(2) = 16 - 3(8) + 4(4) - 5 = 3$. Thus $N \approx 12(.01) + 3 = 3.12$. The exact value of N works out to be 3.12100501.

EXERCISES 3.5, page 119

1. By (3.38), $f'(x) = 3(x^2 - 3x + 8)^2 \cdot D_x(x^2 - 3x + 8) = 3(x^2 - 3x + 8)^2 (2x - 3).$

4. $k'(x) = -3(5x^2 - 2x + 1)^{-4} \cdot D_x(5x^2 - 2x + 1) = -3(5x^2 - 2x + 1)^{-4} (10x - 2).$

7. $D_x(8x^3 - 2x^2 + x - 7)^5 = 5(8x^3 - 2x^2 + x - 7)^4 \cdot D_x(8x^3 - 2x^2 + x - 7) =$
   $5(8x^3 - 2x^2 + x - 7)^4(24x^2 - 4x + 1).$

10. $K'(x) = (-1)(3x^2 - 5x + 7)^{-2} D_x(3x^2 - 5x + 7) = -(3x^2 - 5x + 7)^{-2} (6x - 5).$

13. $N'(x) = (6x - 7)^3 D_x(8x^2 + 9)^2 + (8x^2 + 9)^2 D_x(6x - 7)^3$

    $= (6x - 7)^3 \cdot 2(8x^2 + 9) D_x(8x^2 + 9) + (8x^2 + 9)^2 \cdot 3(6x - 7)^2 D_x(6x-7)$

    $= (6x - 7)^3 \cdot 2(8x^2 + 9)(16x) + (8x^2 + 9)^2 \cdot 3(6x - 7)^2 \cdot 6$

    $= (6x - 7)^2(8x^2 + 9)[32x(6x - 7) + 18(8x^2 + 9)]$

    $= (6x - 7)^2(8x^2 + 9)(336x^2 - 224x + 162).$

16. $S'(t) = 3(\frac{3t+4}{6t-7})^2 D_t(\frac{3t+4}{6t-7}) = 3(\frac{3t+4}{6t-7})^2 [\frac{(6t-7)(3)-(3t+4)(6)}{(6t-7)^2}]$

    $= 3(\frac{3t+4}{6t-7})^2 \frac{-45}{(6t-7)^2} = \frac{-135(3t+4)^2}{(6t-7)^4}.$

19. $f'(x) = 2(\frac{3x^2-5}{2x^2+7})D_x(\frac{3x^2-5}{2x^2+7}) = 2(\frac{3x^2-5}{2x^2+7})[\frac{(2x^2+7)(6x)-(3x^2-5)(4x)}{(2x^2+7)^2}]$

    $= 2(\frac{3x^2-5}{2x^2+7})(\frac{62x}{(2x^2+7)^2}) = \frac{124x(3x^2-5)}{(2x^2+7)^3}.$

22. $F'(v) = (-3)(v^{-1}-2v^{-2})^{-4}D_v(v^{-1}-2v^{-2}) = -3(v^{-1}-2v^{-2})^{-4}(-v^{-2}+4v^{-3}).$

25. $F'(t) = [2t(2t+1)^2]D_t(2t+3)^3 + (2t+3)^3 D_t[2t(2t+1)^2]$

    $= [2t(2t+1)^2]3(2t+3)^2 \cdot 2 + (2t+3)^3[2t \cdot 2(2t+1) \cdot 2 + (2t+1)^2 \cdot 2]$
       (by the product and chain rules)
    $= 2(2t+1)(2t+3)^2(24t^2 + 26t + 3).$

28. $y' = 5(x + 1/x)^4 D_x(x + 1/x) = 5(x + 1/x)^4(1 - 1/x^2)$

    (a) $x = 1 \implies y' = 5(2)^4(1-1) = 0$. Thus the tangent line is $y = 32$.
    (b) The tangent line is horizontal if $y'(x) = 0$. Thus the equation
       $5(x + 1/x)^4(1 - 1/x^2) = 0$ or, after multiplying by $x^6/5$,

$(x^2+1)^4(x^2-1) = 0$ with solutions $x = \pm 1$. The points are: (1,32), (-1,-32).

31. $dy = y'\Delta x = 10(x^4 - 3x^2 + 1)^9(4x^3 - 6x)\Delta x$. Since x changes from 1 to 1.01,

we take $x = 1$, $\Delta x = .01$. Then $\Delta y \approx dy = 10(1 - 3 + 1)^9 \cdot (4 - 6)(.01) =$

$10(-1)^9(-2)(.01) = 20(.01) = 0.2$.

34. If $v = F(u)$ and $u = G(t)$, then $v = F(G(t))$ and $\dfrac{dv}{dt} = F'(G(t))G'(t) = \dfrac{dv}{du} \cdot \dfrac{du}{dt}$.

36. Using #34 with $v = (u^4 + 2u^2 + 1)^3$, $u = 4t^2$, we get $\dfrac{dv}{dt} = 3(u^4 + 2u^2 + 1)^2 \cdot$

$(4u^3 + 4u) \cdot 8t$.

37. Let $y = f(g(x))$. Here, $x = 1$, $x+\Delta x = 0.99$ so that $\Delta x = -.01$. $\Delta y \approx dy =$

$f'(g(x))g'(x)\Delta x = f'(g(1))g'(1)\Delta x$. Now, $g(1) = 1$, and $g'(x) = 3x^2 - 6x + 2$

$\Longrightarrow g'(1) = -1$. Next, $f'(x) = 4x^3 - 9x^2 + 3 \Longrightarrow f'(g(1)) = f'(1) = -2$. Sub-

stituting: $\Delta y \approx (-2)(-1)(-.01) = -.02$.

40. $r(x) = s(t(x)) \Longrightarrow r'(x) = s'(t(x))t'(x) \Longrightarrow r'(0) = s'(t(0))t'(0) =$

$s'(0)t'(0)$ since $t(0) = 0$. Using $r'(0) = 2$ and $s'(0) = -3$, we get $2 =$

$-3t'(0)$ so that $t'(0) = -2/3$.

43. $f'(x_1)$ is the slope of the tangent line at $P \Rightarrow -1/f'(x_1)$ is the slope of the

normal line, perpendicular to the tangent.

46. $f'(x) = -3x^2 \Longrightarrow f'(1) = -3 \Longrightarrow -1/f'(1) = 1/3$, the slope of the normal. The

equation is $y - 7 = (1/3)(x-1)$ or $x - 3y + 20 = 0$.

49. Let (x_1, y_1) be the point of intersection of the graph and the normal line of

slope 4. Then the tangent line has slope $-1/4$ there. $y = (8x+3)^{-2} \Longrightarrow y'$

$= -2(8x+3)^{-3}(8)$. Now, $y'(x_1) = -1/4 \Longleftrightarrow -16(8x_1+3)^{-3} = -1/4 \Longleftrightarrow (8x_1+3)^3$

$= 64 \Longleftrightarrow 8x_1 + 3 = 4 \Longleftrightarrow x_1 = 1/8$. Thus $y_1 = (8(1/8)+3)^{-2} = 4^{-2} = 1/16$.

Now using $m = 4$, the equation is: $y - 1/16 = 4(x-1/8)$.

52. $D_x f(g(x)) = f'(g(x))g'(x) = f(g(x))g'(x)$ for this special function.

EXERCISES 3.6, page 125

1. The object in #1-8 is to solve the given equation for y in terms of x. Thus

$3x - 2y + 4 = 2x^2 + 3y - 7x \Longrightarrow -2x^2 + 10x + 4 = 5y$ and $y = (-2/5)x^2 + 2x +$

$(4/5)$ is one implicit function defined by the equation. Its domain is $\mathbb{R}$.

4. $-3x^2 + 4y^2 = -12 \Longleftrightarrow y^2 = (3x^2 - 12)/4$. So, $y = (1/2)\sqrt{3x^2 - 12}$ is one

function defined by the equation. The domain is $\{x: 3x^2 - 12 \geq 0\} =$

$(-\infty, -2] \cup [2, \infty)$.

7. $\sqrt{x} + \sqrt{y} = 1 \Longleftrightarrow \sqrt{y} = 1 - \sqrt{x} \Longrightarrow y = 1 - 2\sqrt{x} + x$. The domain is tricky. Ob-

viously $x \geq 0$. But from the 2nd equation $\sqrt{y} = 1 - \sqrt{x} \geq 0$ or $\sqrt{x} \leq 1$, which gives $0 \leq x \leq 1$ as the domain.

10. $4x^3 - 2y^3 = x \implies 12x^2 - 6y^2y' = 1 \implies y' = (12x^2-1)/6y^2$.

13. $5x^2 - xy - 4y^2 = 0 \implies 10x - (xy'+y) - 8yy' = 0$ (The middle term is $D_x(xy)$)

$\implies (-x-8y)y' = y - 10x \implies y' = (10x-y)/(x+8y)$.

16. $x = y + 2y^2 + 3y^3 \implies 1 = y' + 4yy' + 9y^2y' \implies y' = 1/(1 + 4y + 9y^2)$.

19. $(y^2-9)^4 = (4x^2 + 3x - 1)^2 \implies 4(y^2-9)^3 D_x(y^2-9) = 2(4x^2 + 3x - 1) \cdot$

$D_x(4x^2+3x-1) \implies 4(y^2-9)^3 \, 2yy' = 2(4x^2 + 3x - 1)(8x + 3) \implies y' = $
$(4x^2 + 3x - 1)(8x + 3)/4y(y^2-9)^3$.

22. $y^2 - 4x^2 = 5 \implies 2yy' - 8x = 0 \implies y' = 4x/y$. At $P(-1,3)$, $y' = -4/3 \implies$
$y-3 = (-4/3)(x+1)$ is the tangent line equation.

28. $x^2 + y^2 = a^2 \implies 2x + 2yy' = 0 \implies y' = -x/y$. If $P(x_1,y_1)$ is on the circle

and $y_1 \neq 0$, the tangent has slope $y' = -x_1/y_1$ and $m_{OP} = y_1/x_1 = -1/y' \implies$

OP $\perp$ tangent. If $y_1 = 0$, then $x_1=\pm$ a and the tangent is vertical while OP

is on the x-axis, horizontal.

EXERCISES 3.7, page 128

1. $f(x) = \sqrt[3]{x^2} + 4\sqrt{x^3} = x^{2/3} + 4x^{3/2} \implies f'(x) = \frac{2}{3}x^{-1/3} + 4(\frac{3}{2})x^{1/2} = \frac{2}{3\sqrt[3]{x}} + 6\sqrt{x}$.

4. $D_z[(2z^2 - 9z + 8)^{-2/3}] = (-2/3)(2z^2 - 9z + 8)^{-5/3} \quad D_z(2z^2 - 9z + 8) = $
$(-2/3)(2z^2 - 9z + 8)^{-5/3}(4z - 9)$.

7. $D_x(\sqrt{2x}) = \sqrt{2}D_x(x^{1/2}) = (\sqrt{2}/2)x^{-1/2} = 1/\sqrt{2x}$.

10. $f(t) = \sqrt[3]{t^2} - 1/\sqrt{t^3} = t^{2/3} - t^{-3/2} \implies f'(t) = \frac{2}{3}t^{-1/3} + \frac{3}{2}t^{-5/2} = 2/3\sqrt[3]{t} + 3/2\sqrt{t^5}$.

13. $D_x(\sqrt{4x^2 - 7x + 4}) = D_x(4x^2 - 7x + 4)^{1/2} = \frac{1}{2}(4x^2 - 7x + 4)^{-1/2}D_x(4x^2 - 7x + 4)$
$= (8x - 7)/2\sqrt{4x^2 - 7x + 4}$.

16. $D_v(1/\sqrt{v^4 + 7v^2})^3 = D_v(v^4 + 7v^2)^{-3/2} = -\frac{3}{2}(v^4 + 7v^3)^{-5/2}(4v^3 + 14v)$.

19. $D_s((s^2+9)^{1/4}(4s+5)^4) = (s^2+9)^{1/4}D_s(4s+5)^4 + (4s+5)^4 D_s(s^2+9)^{1/4}$

$= (s^2+9)^{1/4}4(4s+5)^3(4) + (4s+5)^4(1/4)(s^2+9)^{-3/4}(2s)$
$= 16\sqrt[4]{s^2+9} (4s+5)^3 + s(4s+5)^4/2\sqrt[4]{(s^2+9)^3}$.

22. $D_w(w^3(9w+1)^5)^{1/2} = (1/2)[w^3(9w+1)^5]^{-1/2}D_w[w^3(9w+1)^5] = (1/2)[w^3(9w+1)^5]^{-1/2} \cdot$
$[w^3 \cdot 5(9w+1)^4 \cdot 9 + (9w+1)^5 \cdot 3w^2]$ which, believe it or not, can be simplified to

$(3/2)(24w+1)\sqrt{w(9w+1)^3}$.

25. $D_x(7x+(x^2+3)^{1/2})^6 = 6(7x+\sqrt{x^2+3})^5 D_x(7x+(x^2+3)^{1/2}) = 6(7x+\sqrt{x^2+3})^5(7+x/\sqrt{x^2+3})$.

28. $y = (5x-8)^{1/3} \implies y' = (1/3)(5x-8)^{-2/3} \cdot 5$. At $P(7,3)$, $y' = (1/3)(35-8)^{-2/3} \cdot 5$

$= (5/3)/27^{2/3} = 5/27$. Thus tangent line: $y-3 = (5/27)(x-7)$.

31. $y = (2x-4)^{1/2} \implies y' = (1/2)(2x-4)^{-1/2}(2) = 1/\sqrt{2x-4}$. Thus the slope of the tangent line at $(a,\sqrt{2a-4})$ is $1/\sqrt{2a-4}$ and the equation is $y - \sqrt{2a-4} = (1/\sqrt{2a-4})(x-a)$. Since the desired line is to pass through the origin, $x = y = 0$ must satisfy this equation. Thus $-\sqrt{2a-4} = -a/\sqrt{2a-4} \implies 2a-4 = a \iff a = 4$. Thus the point is $(4,\sqrt{8-4}) = (4,2)$.

34. $x^{2/3} + y^{2/3} = 4 \iff \frac{2}{3}x^{-1/3} + \frac{2}{3}y^{-1/3}y' = 0 \implies y^{-1/3}y' = -x^{-1/3} \implies y'$

$= -\frac{x^{-1/3}}{y^{-1/3}} = -(\frac{y}{x})^{1/3}$.

37. $3x^2 + \sqrt[3]{xy} = 2y^2 + 20 \implies 3x^2 + x^{1/3}y^{1/3} = 2y^2 + 20 \implies 6x + x^{1/3}\frac{1}{3}y^{-2/3}y' + \frac{1}{3}x^{-2/3}y^{1/3} = 4yy'$ (having used the product rule for $D_x(x^{1/3}y^{1/3})$). Multiplying by 3 and regrouping: $(x^{1/3}y^{-2/3} - 12y)y' = -18x - x^{-2/3}y^{1/3} \implies$

$y' = \frac{18x + x^{-2/3}y^{1/3}}{12y - x^{1/3}y^{-2/3}} = \frac{18x^{5/3}y^{2/3} + y}{12x^{2/3}y^{5/3} - x}$ (having multiplied and divided by $x^{2/3}y^{2/3}$).

40. $dy = D_x y \Delta x = D_x[(x^2+1)^{-1/2}]\Delta x = (-1/2)(x^2+1)^{-3/2}D_x(x^2+1)\Delta x = -\frac{x}{\sqrt{(x^2+1)^3}}\Delta x$.

43. $dy = D_x(4x^2+9)^{3/2}\Delta x = (3/2)(4x^2+9)^{1/2}D_x(4x^2+9)\Delta x = (3/2)\sqrt{4x^2+9}\,8x\Delta x$. With $x = 2$, $x+\Delta x = 1.998$, $\Delta x = -.002$. $\Delta y \approx dy = (3/2)\sqrt{16+9}\,(16)(-.002) = -120(.002) = -.24$.

46. Write $T = K\ell^{1/2}$ where $K = 2\pi/\sqrt{g}$. We seek $\Delta\ell$ so that $\frac{\Delta T}{T}(100) = 1$ (i.e., a 1% increase in T) or $\frac{\Delta T}{T} = .01$. Now, $\frac{\Delta T}{T} \approx \frac{dT}{T} = \frac{(1/2)K\ell^{-1/2}\Delta\ell}{K\ell^{1/2}} = \frac{\Delta\ell}{2\ell} = .01$ if $\frac{\Delta\ell}{\ell} = .02$ or $\Delta\ell = .02\ell = \frac{\ell}{50}$. (i.e., ℓ must increase by 2%.)

EXERCISES 3.8, page 131

1. See Example 1.

2. $g'(x) = 24x^7 - 10x^4$, $g''(x) = 168x^6 - 40x^3$.

4. $F'(t) = (3/2)t^{1/2} - t^{-1/2} - 2t^{-3/2}$, $F''(t) = (3/4)t^{-1/2} + (1/2)t^{-3/2} + 3t^{-5/2}$.

7. $k'(r) = 5(4r+7)^4 D_r(4r+7) = 20(4r+7)^4$.

$$k''(r) = 20(4)(4r+7)^3 D_r(4r+7) = 80(4r+7)^3 \cdot 4.$$

10. $h(x) = 1 \implies h'(x) = 0 \implies h''(x) = 0.$

13. $D_x y = \dfrac{(3x+1)D_x(2x-3)-(2x-3)D_x(3x+1)}{(3x+1)^2} = \dfrac{(3x+1)(2)-(2x-3)(3)}{(3x+1)^2} = 11(3x+1)^{-2}.$

   $D_x^2 y = -2(11)(3x+1)^{-3}D_x(3x+1) = -66(3x+1)^{-3}$

   $D_x^3 y = (-3)(-66)(3x+1)^{-4}D_x(3x+1) = 198(3)(3x+1)^{-4}.$

16. $D_x y = 4(3x+1)^3 D_x(3x+1) = 12(3x+1)^3.$ $D_x^2 y = 36(3x+1)^2 D_x(3x+1) = 108(3x+1)^2.$

   $D_x^3 y = 216(3x+1) \cdot D_x(3x+1) = 648(3x+1).$

19. $x^2 - 3xy + y^2 = 4 \implies 2x-3xy'-3y+2yy'=0 \implies y'=(3y-2x)/(2y-3x) \implies y'' = $

   $\dfrac{(2y-3x)(3y'-2) - (3y-2x)(2y'-3)}{(2y-3x)^2}$. Substituting for y' (from the 1st line),

   the numerator of y'' becomes: $3(3y-2x)-2(2y-3x)-2\dfrac{(3y-2x)^2}{(2y-3x)} + 3(3y-2x)$

   $= (14y-6x) - 2(3y-2x)^2/(2y-3x) = [(14y-6x)(2y-3x) - 2(3y-2x)^2]/(2y-3x)$

   $= [10y^2 - 30xy + 10x^2]/(2y-3x)$, which when substituted into the y'' equation

   yields the answer given in the text.

22. $f(x) = (x^2-1)^3 = x^6 - 3x^4 + 3x^2 - 1 \implies f'(x) = 6x^5 - 12x^3 + 6x \implies f''(x) = $

   $30x^4 - 36x^2 + 6 \implies f'''(x) = 120x^3 - 72x \implies f^{(4)}(x) = 360x^2 - 72 \implies$

   $f^{(5)}(x) = 720x \implies f^{(6)}(x) = 720.$

28. Let $g(x) = f'(x) = 4x^3 - 3x^2 - 12x + 7.$ A tangent line to the graph of g
   has slope $g'(x)$ or $f''(x) = 12x^2 - 6x - 12.$ At $x = 2$, $g'(2) = f''(2) = $
   $12(4) - 6(2) - 12 = 24$, and the equation of the tangent line at $(2,3)$ is
   $y-3 = 24(x-2).$

31. $y = f(g(x)) \implies D_x y = f'(g(x))g'(x).$ Using the product rule: $D_x^2 y = $

   $f'(g(x))D_x g'(x) + [D_x f'(g(x))]g'(x) = f'(g(x))g''(x) + [f''(g(x))g'(x)]g'(x).$

EXERCISES 3.9 (Review), page 132

1. $f'(x) = \lim\limits_{h \to 0} \dfrac{1}{h}\Big(\dfrac{4}{3(x+h)^2+2} - \dfrac{4}{3x^2+2}\Big) = \lim\limits_{h \to 0} \dfrac{4}{h}\dfrac{-(6xh-3h^2)}{(3(x+h)^2+2)(3x^2+2)} = \dfrac{-24x}{(3x^2+2)^2}.$

4. $D_x(x^4-x^2+1)^{-1} = (-1)(x^4-x^2+1)^{-2}D_x(x^4-x^2+1) = -(4x^3-2x)/(x^4-x^2+1)^2.$

7. $D_z(7z^2-4z+3)^{1/3} = (1/3)(7z^2-4z+3)^{-2/3}D_z(7z^2-4z+3) = (14z-4)/3\sqrt[3]{(7z^2-4z+3)^2}.$

10. $D_x((1/6)(3x^2-1)^4) = (4/6)(3x^2-1)^3(6x) = 4x(3x^2-1)^3.$

13. $D_x(\sqrt[5]{(3x+2)^4}) = D_x(3x+2)^{4/5} = (4/5)(3x+2)^{-1/5}(3).$

16. Multiply out to get $g(w) = (w^2-4w+3)/(w^2+4w+3).$ Then

$g'(w) = \dfrac{(w^2+4w+3)(2w-4) - (w^2-4w+3)(2w+4)}{(w+1)^2(w+3)^2} = (8w^2-24)/(w+1)^2(w+3)^2.$

19. $g'(y) = (7y-2)^{-2}D_y(2y+1)^{2/3} + (2y+1)^{2/3}D_y(7y-2)^{-2} = (7y-2)^{-2}(2/3)(2y+1)^{-1/3}(2)$

  $+ (2y+1)^{2/3}(-2)(7y-2)^{-3}(7) = (7y-2)^{-3}(2y+1)^{-1/3}[(4/3)(7y-2) - 14(2y+1)]$
  which reduces to the answer given.

22. $D_t(t^6-t^{-6})^6 = 6(t^6-t^{-6})^5D_t(t^6-t^{-6}) = 6(t^6-t^{-6})^5(6t^5+6t^{-7}).$

25. $f'(x) = [(3x+2)^{1/2}D_x(2x+3)^{1/3} - (2x+3)^{1/3}D_x(3x+2)^{1/2}]/(3x+2)$

  $= \dfrac{(3x+2)^{1/2}(\frac{1}{3})(2x+3)^{-2/3}(2)-(2x+3)^{1/3}(\frac{1}{2})(3x+2)^{-1/2}(3)}{3x+2}$

  $= (3x+2)^{-1/2}(2x+3)^{-2/3}[(2/3)(3x+2)-(3/2)(2x+3)]/(3x+2)$

  which may be reduced to $-(x + 19/6)/(3x+2)^{3/2}(2x+3)^{2/3}.$

28. $F'(t) = [(t^2+2)(10t) - (5t^2-7)(2t)]/(t^2+2)^2 = 34t/(t^2+2)^2.$

31. $f'(w) = \dfrac{1}{2}(\dfrac{2w+5}{7w-9})^{-1/2}[\dfrac{(7w-9)(2)-(2w+5)(7)}{(7w-9)^2}] = \dfrac{1}{2}(\dfrac{2w+5}{7w-9})^{-1/2}[\dfrac{-53}{(7w-9)^2}],$ which
  reduces to the given answer.

34. $3x^2-xy^2+y^{-1} = 1 \implies 6x-2xyy'-y^2-y^{-2}y' = 0 \implies y' = (6x-y^2)/(2xy+y^{-2}).$

37. $y = 2x-4x^{-1/2} \implies y' = 2 + 2x^{-3/2}.$ At $x = 4,$ $y' = 2 + 2(2^{-3}) = 2 + 2/8 =$
  $9/4.$ Thus, the tangent line is $y-6 = (9/4)(x-4).$

40. $y = 5x^3+4x^{1/2} \implies y' = 15x^2+2x^{-1/2} \implies y'' = 30x - x^{-3/2} \implies y''' = 30 +$
  $(3/2)x^{-5/2}.$

43. $x^2 + 4xy - y^2 = 8 \implies 2x + 4xy' + 4y - 2yy' = 0 \implies y' = (2x+4y)/(2y-4x)$
  $= (x+2y)/(y-2x) \implies y'' = [(y-2x)(1+2y')-(x+2y)(y'-2)]/(y-2x)^2$
  $= 5(y - xy')/(y - 2x)^2.$ Substituting for y' from the 2nd line and simplifying yields the given answer.

46. $y' = [(x^2+1)(5) - 5x(2x)]/(x^2+1)^2 = 5(1-x^2)/(x^2+1)^2.$ Thus $dy =$
  $[5(1-x^2)/(x^2+1)^2]\Delta x.$ If x changes from 2 to 1.98, we select $x = 2,$ $\Delta x =$
  $-.02$ and $\Delta y \approx dy = (5(1-4)/5^2)(-.02) = (-3/5)(-.02) = (.6)(.02) = 0.012.$
  Exactly, $\Delta y = f(1.98)-f(2) = \dfrac{9.9}{4.9204} - \dfrac{10}{5} \approx 0.0120315.$

49. Let $h(x) = g(f(x)).$ If x changes from -1 to $-1.01,$ then select $x = -1,$
  $\Delta x = -.01$ and $\Delta y \approx dy = h'(x)\Delta x = g'(f(x))f'(x)\Delta x = g'(f(-1))f'(-1)\Delta x.$ Now,
  $f'(x) = 6x^2 + 2x - 1,$ $f'(-1) = 6-2-1 = 3,$ $g'(x) = 5x^4 + 12x^2 + 2,$ $f(-1) =$
  $-2+1+1+1 = 1,$ $g'(f(-1)) = g'(1) = 5+12+2 = 19.$ Thus $\Delta y \approx (19)(3)(-.01) =$
  $-0.57.$

CHAPTER 4

APPLICATIONS OF THE DERIVATIVE

EXERCISES 4.1, page 142

1. $f'(x) = -12x - 6x^2 = -6x(x+2)$. Since $f'(x)$ exists for all x, the only critical numbers are solutions of $f'(x) = 0$, or $-6x(x+2) = 0 \Longrightarrow x = 0$, $x = -2$, both of which are in the given interval $[-3,1]$. Now we calculate the values of $f(x)$ at $x = 0$ and $x = -2$, the critical numbers, and at $x = -3$ and $x = 1$, the end points of the interval. The absolute maximum of f on $[-3,1]$ is the largest of these values, and the absolute minimum of f is the smallest. $f(x) = 5 - 6x^2 - 2x^3 \Longrightarrow f(0) = 5$, $f(-2) = 5 - 6(4) - 2(-8) = -3$, $f(-3) = 5 - 6(9) - 2(-27) = 5$, $f(1) = 5 - 6 - 2 = -3$. Thus the maximum is 5 attained at $x = 0$ and -3; the minimum is -3 attained at $x = 1$ and -2.

4. $f'(x) = 4x^3 - 10x = 2x(2x^2-5)$, which exists for all x. Thus the only critical numbers are solutions of $f'(x) = 0$, or $2x(2x^2 - 5) = 0 \Longrightarrow x = 0$ and $x = \pm\sqrt{5/2}$. Of these, $-\sqrt{5/2}$ is not in the given interval, $[0,2]$, but 0 and $\sqrt{5/2} \approx 1.6$ are. Again, we calculate $f(x)$ at the critical numbers in the interval and at the end points. ($x = 0$ is both, but we consider it only an end point since, strictly speaking, $f'(0)$ does not exist--only the right-hand derivative does exist.) $f(0) = 4$, $f(2) = 16 - 5(4) + 4 = 0$, $f(\sqrt{5/2}) = (\sqrt{5/2})^4 - 5(\sqrt{5/2})^2 + 4 = 25/4 - 25/2 + 4 = -9/4$. Thus, the maximum is $f(0) = 4$, and the minimum is $f(\sqrt{5/2}) = -9/4$.

6. For $f(x) = |x|$, $f'(0)$ does not exist by Example 3, Sec. 3.2, but $f'(x) = 1$ if $x > 0$, $f'(x) = -1$ if $x < 0$. Thus $x = 0$ is the only critical point, and the graph has no tangent line at $(0,0)$. Since $f(0) = 0$, f has a local minimum at $x = 0$ because $f(x) = |x| \geq 0 = f(0)$ for all x in any interval (a,b) containing 0.

10. $g'(x) = 2$, exists for all x and is never 0. Thus, no critical numbers.

13. $F'(w) = 4w^3 - 32 = 4(w^3-8) = 0$ only if $w = 2$, the only critical number.

16. $M'(x) = (2x-1)/3(x^2-x-2)^{1/3} = (2x-1)/3\sqrt[3]{(x-2)(x+1)}$. Thus $M'(x) = 0$ if $x = 1/2$, and $M'(x)$ fails to exist at $x = -1$ and $x = 2$.

19. $G'(x) = (-2x^2 + 6x - 18)/(x^2-9)^2$ fails to exist at $x = \pm3$, but neither of these are in the domain of G. Setting $G'(x) = 0$, we obtain no real solutions since $b^2 - 4ac = 36 - 144 < 0$. Thus, no critical numbers.

22. If $f(x) = k$ for all x in (a,b) and if $c \in (a,b)$, then $f(x) = f(c)$ for all x there, and so $f(x) \geq f(c)$ and $f(x) \leq f(c)$ are both true.

28. If $f(x)$ is a polynomial of degree n, then $f'(x)$ has degree $n-1$ and can have at most $n-1$ distinct zeros. Since $f'(x)$, as a polynomial, exists everywhere, these are the only possible critical points.

EXERCISES 4.2, page 146

1. f'(0) doesn't exist by Example 3, Section 3.2.

4. If [a] = n and b-a $\geq$ 1, then b $\geq$ a+1 and [b] $\geq$ n+1. Thus f(b) - f(a) = [b] -
    [a] $\geq$ 1. Trying to solve f(b) - f(a) = f'(c)(b-a) we get f'(c) =
    $\frac{f(b)-f(a)}{b-a} \geq \frac{1}{b-a}$. This equation has no solutions since f'(x) = 0 if x is not
    an integer and f'(x) does not exist if x is an integer. (See #22, Section
    3.2.) There is no contradiction since b-a $\geq$ 1 $\Longrightarrow$ [a,b] contains at least one
    integer at which f is neither continuous nor differentiable.

7. As a polynomial function, f is continuous and differentiable everywhere.
    Rolle's Theorem applies since f(3) = f(-3) = 118. f'(x) = $4x^3$ + 8x = $4x(x^2+2)$
    = 0 only at x = c = 0, the desired solution.

10. As a polynomial function, f is continuous and differentiable everywhere.
    With a = 1, b = 3 we obtain: f(3) - f(1) = f'(c)(3-1) $\Longleftrightarrow$ 37 - 3 =
    (10c-3) $\cdot$ 2 $\Longleftrightarrow$ 17 = 10c - 3 $\Longleftrightarrow$ c = 2.

13. f is continuous on [-8,8] but is not differentiable on (-8,8) since f'(0) does
    not exist. ($\lim_{h \to 0} \frac{f(h) - f(0)}{h} = \lim_{h \to 0} \frac{h^{2/3}}{h} = \lim_{h \to 0} \frac{1}{h^{1/3}}$ which does not exist.)

16. f(x) = 1 - $3x^{1/3}$ is continuous on [-8,-1] and differentiable on (-8,-1).
    (f'(0) does not exist, but 0 $\notin$ (-8,-1).) f(-1) - f(-8) = f'(c)(-1 -(-8)) $\Longleftrightarrow$
    4-7 = $(-c^{-2/3})(7)$ $\Longleftrightarrow$ $c^{-2/3}$ = 3/7 $\Longleftrightarrow$ $c^{2/3}$ = 7/3 $\Longleftrightarrow$ $c^2 = (7/3)^3$ $\Longleftrightarrow$
    c = $\pm \sqrt{(7/3)^3}$. The negative root must be chosen for c to be in (-8,-1). Thus
    c = $-\sqrt{(7/3)^3}$ $\approx$ -3.6.

20. f'(x) is a linear polynomial (i.e. of degree one). Thus f(b) - f(a) =
    f'(c)(b-a) is a linear equation in the unknown c.

EXERCISES 4.3, page 153

1. The critical numbers are solutions of f'(x) = -7 - 8x = 0 or x = -7/8. For
    x < -7/8, f'(x) > 0 and f is increasing; for x > -7/8, f'(x) < 0 and f is
    decreasing. Thus f(-7/8) is a local maximum.

4. f'(x) = $3x^2$ - 2x - 40 = (3x+10)(x-4) = 0 if
    x = -10/3 and x = 4.
    Tabulating our work:

Interval	(3x+10)	(x-4)	f'	f
$(-\infty,-10/3)$	$-$	$-$	$+$	increasing
$(-10/3,4)$	$+$	$-$	$-$	decreasing
$(4,\infty)$	$+$	$+$	$+$	increasing

Thus a local maximum $f(-10/3) = 2516/27$, and a local minimum $f(4) = -104$.

7. $f'(x) = \frac{4}{3}(x^{1/3} + x^{-2/3}) = 4(x+1)/3x^{2/3}$. Thus critical numbers, $x = -1$ and $x = 0$.

Interval	(x+1)	$x^{2/3}$	f'	f
$(-\infty,-1)$	$-$	$+$	$-$	decreasing
$(-1,0)$	$+$	$+$	$+$	increasing
$(0,\infty)$	$+$	$+$	$+$	increasing

Thus a local minimum $f(-1) = -3$. (No extremum at $x = 0$. Only a vertical tangent.)

10. $f'(x) = x(-2x/2\sqrt{4-x^2}) + \sqrt{4-x^2} = 2(2-x^2)/\sqrt{4-x^2} = 0$ at $x = \pm\sqrt{2}$.

Interval	$2-x^2$	f'	f
$(-2,-\sqrt{2})$	$-$	$-$	decreasing
$(-\sqrt{2},\sqrt{2})$	$+$	$+$	increasing
$(\sqrt{2},2)$	$-$	$-$	decreasing

Thus a local maximum $f(\sqrt{2}) = 2$ and a local minimum $f(-\sqrt{2}) = -2$.

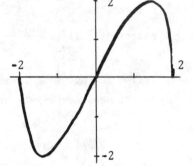

13. $f'(x) = 3x^2 - 3/x^2 = 3(x^4-1)/x^2 = 3(x^2+1)(x^2-1)/x^2 = 0$ at $x = \pm 1$. ($x = 0$ is not a critical number even though $f'(0)$ doesn't exist since 0 is not in the domain of f. $f(x)$ becomes arbitrarily large and positive as $x \to 0^+$, large and negative as $x \to 0^-$.) Tabulating as before:

Interval	$3(x^2+1)/x^2$	x^2-1	f'	f
$(-\infty,-1)$	$+$	$+$	$+$	increasing
$(-1,0)$	$+$	$-$	$-$	decreasing
$(0,1)$	$+$	$-$	$-$	decreasing
$(1,\infty)$	$+$	$+$	$+$	increasing

Thus $f(-1) = -4$ is a local maximum, and $f(1) = 4$ is a local minimum.

16. $f'(x) = 4(x^2-10x)^3 D_x(x^2-10x) = 4(x^2-10x)^3(2x-10) = 8[x(x-10)]^3(x-5) = 8x^3(x-10)^3(x-5) = 0$ at $x = 0,5,10$.

Interval	$8x^3$	$(x-5)$	$(x-10)^3$	f'	f
$(-\infty,0)$	−	−	−	−	decreasing
$(0,5)$	+	−	−	+	increasing
$(5,10)$	+	+	−	−	decreasing
$(10,\infty)$	+	+	+	+	increasing

Thus $f(0) = 0$ and $f(10) = 0$ are local minima, and $f(5) = 25^4$ is a local maximum.

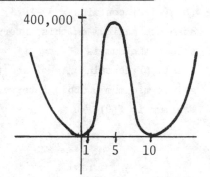

19. $f'(x) = (x-2)^3 \cdot 4(x+1)^3 + (x+1)^4 \cdot 3(x-2)^2 = (x-2)^2(x+1)^3(7x-5) = 0$ at $x = 2$, -1, and $5/7$.

Interval	$(x-2)^2$	$(x+1)^3$	$(7x-5)$	f'	f
$(-\infty,-1)$	+	−	−	+	increasing
$(-1,5/7)$	+	+	−	−	decreasing
$(5/7,2)$	+	+	+	+	increasing
$(2,\infty)$	+	+	+	+	increasing

Thus local maximum $f(-1) = 0$ and a local minimum $f(5/7) = (-9/7)^3(12/7)^4 \approx$ -18.4. (No extremum at $x = 2$, only a horizontal tangent.)

22. $f'(x) = [(x-1)(2x) - (x^2+3)]/(x-1)^2 = (x-3)(x+1)/(x-1)^2 = 0$ at $x = -1$ and 3, the only critical numbers.

Interval	$(x-3)$	$(x+1)$	$(x-1)^2$	f'	f
$(-\infty,-1)$	−	−	+	+	increasing
$(-1,3)$	−	+	+	−	decreasing
$(3,\infty)$	+	+	+	+	increasing

There is a local maximum $f(-1) = -2$ and a local minimum $f(3) = 6$.

NOTE: Solutions to #23 and #26, which refer to #1 and #4 above, will be included rather than #25.

23. Since f has only one local maximum, $f(-7/8) = 8\frac{1}{16}$ this will be the absolute maximum on any interval containing $-7/8$. The absolute minimum will be at an end point. (a) On $[-1,1]$, $f(-1) = 8$, $f(1) = -6$. Thus, absolute maximum $f(-7/8)$ and absolute minimum $f(1)$.
(b) On $[-4,2]$, $f(-4) = -31$, $f(2) = -25$. Thus, absolute maximum $f(-7/8)$ and absolute minimum $f(-4)$.

(c) On [0,5], f is decreasing. Thus, absolute maximum f(0) and absolute minimum f(5).

26. f here has a local maximum f(-10/3) and a local minimum f(4).

(a) On [-1,1] f is decreasing. Thus, absolute maximum f(-1) = 46 and absolute minimum f(1) = -32.

(b) [-4,2] contains x = -10/3, and the local maximum of f there will be the absolute maximum on this interval. Since f(-4) = 88 and f(2) = -68, the absolute minimum is f(2).

(c) [0,5] contains x = 4, and the local minimum of f there will be the absolute minimum on this interval. Since f(0) = 8 and f(5) = -92, the absolute maximum is f(0).

28.

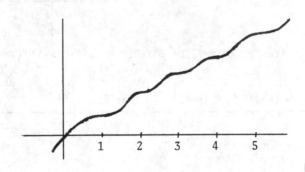

31. $f(x) = ax^3 + bx^2 + cx + d \implies f'(x) = 3ax^2 + 2bx + c$. The given conditions on f mean f(-1) = 2, f(1) = -1, f'(-1) = f'(1) = 0. Substituting these into the f and f' formulas, we get the system of equations:

$$\begin{cases} a + b + c + d = -1 \\ -a + b - c + d = 2 \\ 3a + 2b + c \quad\quad = 0 \\ 3a - 2b + c \quad\quad = 0 \end{cases}$$ Subtract to get 4b = 0 or b = 0.

Using b = 0, we get

$$\begin{cases} a + c + d = -1 \\ -a - c + d = 2 \\ 3a + c \quad\quad = 0 \end{cases}$$ Add to get 2d = 1 or d = 1/2.

Using d = 1/2 in the 1st and 3rd equations, we get

$$\begin{cases} a + c = -3/2 \\ 3a + c = 0 \end{cases}$$ with solution a = 3/4, c = -9/4.

EXERCISES 4.4, page 162

1. $f'(x) = 3x^2 - 4x + 1 = (3x-1)(x-1)$, $f''(x) = 6x-4 = 6(x - 2/3)$. The critical

numbers are x = 1/3, 1 and f"(1/3)= -2. ∴ local maximum f(1/3) = 31/27;

f"(1) = 2 ∴ local minimum f(1) = 1. Since f"(x) > 0 if x > 2/3 and f"(x)

< 0 if x < 2/3, the graph is CU on (-∞,2/3), CD on (2/3,∞), and has a point

of inflection at x = 2/3.

4. $f'(x) = 16x - 8x^3 = 8x(2-x^2)$, $f''(x) = 16-24x^2 =$
$8(2-3x^2)$. The critical numbers are x = 0, $\pm\sqrt{2}$ and
f"(0) = 16 ∴ local minimum f(0) = 0; f"($\pm\sqrt{2}$) =
-32, ∴ local maxima f($\pm\sqrt{2}$) = 8. Next, f"(x) > 0
⟺ $2 - 3x^2 > 0$ ⟺ $x^2 < 2/3$ ⟺ $|x| < \sqrt{2/3}$.
Thus the graph is CU on $(-\sqrt{2/3},\sqrt{2/3})$, CD on
$(-\infty,-\sqrt{2/3})$, and $(\sqrt{2/3},\infty)$ with points of inflection
at x = $\pm\sqrt{2/3}$.

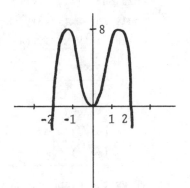

7. $f(x) = (x^2-1)^2 = x^4 - 2x^2 + 1 \Rightarrow f'(x) = 4x^3 - 4x = 4x(x^2-1) \Rightarrow f''(x) =$
$12x^2 - 4 = 4(3x^2 - 1)$. From f'(x) = 0, the critical numbers are x = 0, ± 1.
f"(0) = -4. ∴ local maximum f(0) = 1. f"(± 1) = 8. ∴ local minima f(± 1)
= 0. Next, f"(x) < 0 ⟺ $3x^2 - 1 < 0$ ⟺ $x^2 < 1/3$ ⟺ $|x| < 1/\sqrt{3}$. Thus
the graph is CU on $(-\infty,-1/\sqrt{3})$ and $(1/\sqrt{3},\infty)$, CD on $(-1/\sqrt{3},1/\sqrt{3})$ with points of
inflection at x = $\pm 1/\sqrt{3}$.

10. $f(x) = (x+4)/\sqrt{x} = x^{1/2} + 4x^{-1/2} \Rightarrow f'(x) = \frac{1}{2}x^{-1/2} - 2x^{-3/2} = (x-4)/2x^{3/2} \Rightarrow$
$f''(x) = (x^{3/2} - \frac{3}{2}(x-4)x^{1/2})/2x^3 = \sqrt{x}(12-x)/4x^3$. The only critical number is
x = 4 (not x = 0 since 0 is not in the domain of f), and f"(4) = 1/8. ∴
local minimum f(4) = 4. Noting that f"(x) > 0 if 0 < x < 12, and f"(x) < 0
if x > 12, the graph is CU on (0,12), CD on (12,∞), with a point of inflection at x = 12.

10. (Graph)   16. (Graph)

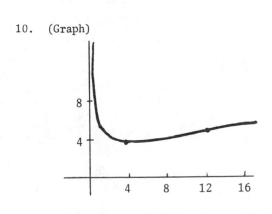

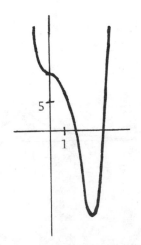

13. $f'(x) = \dfrac{(x^2+1) - 2x^2}{(x^2+1)^2} = \dfrac{1-x^2}{(x^2+1)^2}$. $f''(x) = \dfrac{(x^2+1)^2(-2x) - (1-x^2)2(x^2+1)2x}{(x^2+1)^4}$

$= \dfrac{2x^3 - 6x}{(x^2+1)^3} = \dfrac{2x(x^2-3)}{(x^2+1)^3}$. From $f'(x) = 0$, $x = \pm 1$ are the only critical numbers.

$f''(1) = -4/8$. $\therefore$ a local maximum $f(1) = 1/2$. $f''(-1) = 4/8$. $\therefore$ a local
minimum $f(-1) = -1/2$. Now, $f''(x) = 0$ if $x = 0, \pm\sqrt{3}$. We tabulate our work
only for $2x(x^2-3)$, the numerator of $f''(x)$, since $(x^2+1)^3 > 0$ for all x. The
sign of $f''(x)$ is the same as that of $2x(x^2-3)$.

Interval	$2x$	x^2-3	f''	concavity
$(-\infty, -\sqrt{3})$	-	+	-	CD
$(-\sqrt{3}, 0)$	-	-	+	CU
$(0, \sqrt{3})$	+	-	-	CD
$(\sqrt{3}, \infty)$	+	+	+	CU

Thus the graph has points of inflection at $x = -\sqrt{3}, 0, \sqrt{3}$.

16. $f'(x) = 4x^3 - 12x^2 = 4x^2(x-3) = 0$ at $x = 0, 3$. $f''(x) = 12x^2 - 24x = 12x(x-2)$.
$f''(3) = 36 \implies f(3) = -17$ is a local minimum. $f''(0) = 0$ means that the 2nd
derivative test is not applicable. Trying the 1st derivative test, if $x < 0$
then $f'(x) < 0$, and if $0 < x < 3$ then $f'(x) < 0$ also. Thus f is decreasing
on $(-\infty, 3)$ and has no local extremum at $x = 0$, only a horizontal tangent. Now,
$f''(x) = 0$ at $x = 0, 2$. We tabulate:

Interval	$12x$	$x-2$	f''	concavity
$(-\infty, 0)$	-	-	+	CU
$(0, 2)$	+	-	-	CD
$(2, \infty)$	+	+	+	CU

Thus, PI's at $x = 0$ and $x = 2$. (The graph is next to that of #10 above.)

22.

28. HINT. $f''(x)$ is a linear polynomial.
     Thus $f''(x) = 0$ has only one solution.
     Show $f''(x)$ changes sign there.

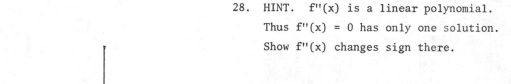

EXERCISES 4.5, page 175

1. $\lim\limits_{x \to \infty} \dfrac{5x^2 - 3x + 1}{2x^2 + 4x - 7} = \lim\limits_{x \to \infty} \dfrac{5 - 3/x + 1/x^2}{2 + 4/x - 7/x^2} = \dfrac{5 - 0 + 0}{2 + 0 - 0} = \dfrac{5}{2}$.

4. $\lim\limits_{x \to -\infty} \dfrac{(3x+4)(x-1)}{(2x+7)(x+2)} = \lim\limits_{x \to -\infty} \dfrac{3x^2 + x - 4}{2x^2 + 11x + 14} = \lim\limits_{x \to -\infty} \dfrac{3 + (1/x) - (4/x^2)}{2 + (11/x) + (14/x^2)} = \dfrac{3}{2}$.

7. $\lim\limits_{x \to \infty} \dfrac{\sqrt{4x+1}}{10-3x} = \lim\limits_{x \to \infty} \dfrac{\sqrt{(4/x) + (1/x^2)}}{10/x - 3} = \dfrac{0}{-3} = 0$. (Since $x \to \infty$, we may assume $x > 0$.

    Thus $x = \sqrt{x^2}$ when dividing numerator and denominator by x.)

10. The procedure here (as in 9 and 12) is to multiply and divide by an expression
    which will clear the radical from the numerator. Recalling that $(a-b)(a+b) = a^2 - b^2$ we have

    $\lim\limits_{x \to \infty} (x - \sqrt{x^2 - 3x}) = \lim\limits_{x \to \infty} \dfrac{(x - \sqrt{x^2 - 3x})(x + \sqrt{x^2 - 3x})}{(x + \sqrt{x^2 - 3x})} = \lim\limits_{x \to \infty} \dfrac{x^2 - (x^2 - 3x)}{x + \sqrt{x^2 - 3x}}$

    $= \lim\limits_{x \to \infty} \dfrac{3x}{x + \sqrt{x^2 - 3x}} = \lim\limits_{x \to \infty} \dfrac{3}{1 + \sqrt{1 - (3/x)}} = \dfrac{3}{1 + \sqrt{1}} = 3/2$.

13. Since $\lim\limits_{x \to \infty} f(x) = \lim\limits_{x \to -\infty} f(x) = \lim\limits_{x \to +\infty} \dfrac{4x^2}{1+x^2} = \lim\limits_{x \to +\infty} \dfrac{4}{1/x^2 + 1} = 4$, the line $y = 4$ is

    a horizontal asymptote. For the graph we need $f'(x) - 8x/(1+x^2)^2$ which is $<$
    0 if $x < 0$ and > 0 if $x > 0$. Thus f is decreasing on $(-\infty, 0]$, increasing on
    $[0, \infty)$, and has a local minimum $f(0) = 0$. Also, $f''(x) = 8(1-3x^2)/(1+x^2)^3 > 0$
    if $|x| < 1/\sqrt{3}$ and < 0 if $|x| > 1/\sqrt{3}$. Thus the graph is CU on $(-1/\sqrt{3}, 1/\sqrt{3})$,
    CD on $(-\infty, -1/\sqrt{3})$ and $(1/\sqrt{3}, \infty)$, with PI's at $x = \pm 1/\sqrt{3}$.

16. If $f(x) = a_n x^n + \ldots + a_0$, $a_n \neq 0$, $n \geq 1$, then $\dfrac{1}{f(x)} = \dfrac{1}{a_n x^n + \ldots + a_0}$

    $= \dfrac{(1/x^n)}{a_n + (a_{n-1}/x) + \ldots + (a_0/x^n)}$, which approaches $0/a_n = 0$ as $x \to \infty$.

19. Given any $\varepsilon > 0$, pick N so that $|g(x) - 0| < \varepsilon$ for $x \geq N$. Since $g(x) > 0$
    this becomes $0 < g(x) < \varepsilon$ for such x. For the same x, $0 < f(x) = |f(x) - 0|$
    $< g(x) < \varepsilon$. Thus $\lim\limits_{x \to \infty} f(x) = 0$.

22. $\lim\limits_{x \to 4^+} \dfrac{5}{4-x} = -\infty$ since $x \to 4^+ \Longrightarrow x > 4$ and $4-x < 0$;

    $\lim\limits_{x \to 4^-} \dfrac{5}{4-x} = \infty$ since now $x < 4$ and $4-x > 0$. The

    graph also uses $\lim\limits_{x \to +\infty} \dfrac{5}{4-x} = 0$.

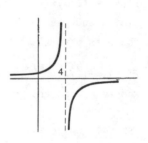

25. If x is near to -8, then the numerator, 3x, is near -24 < 0, whereas the denominator, $(x+8)^2$, is positive and nearly 0. Thus

$$\lim_{x\to-8^+} \frac{3x}{(x+8)^2} = \lim_{x\to-8^-} \frac{3x}{(x+8)^2} = -\infty.$$

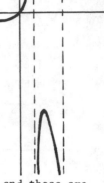

28. $f(x) = 4x(x^2 - 4x + 3) = 4x/(x-1)(x-3)$. For x near
1 or 3, 4x > 0, and the sign of f(x) is that of
(x-1)(x-3). As x → 1⁻, x-1 < 0, x-3 < 0 ⟹ f(x)
> 0 and the limit is ∞. As x → 1⁺ or x → 3⁻, x-1
> 0, x-3 < 0 ⟹ f(x) < 0, and both limits are -∞.
As x → 3⁺, x-1 > 0, x-3 > 0 ⟹ f(x) > 0 and the
limit is ∞. For the graph we need $\lim_{x\to\infty} f(x) = 0$.

Also $f'(x) = 4(3-x^2)/(x^2-4x+3)^2 = 0$ at $x = \pm\sqrt{3}$,
and $f''(x) = 8(x^3-9x+12)/(x^2-4x+3)^3 \Longrightarrow f(\sqrt{3}) \approx$
-7.5 is a local maximum, $f(-\sqrt{3}) \approx -0.53$ is a local
minimum and there is a PI between x = -4 and x = -3.

31. $f(x) = (x+2)(x+1)/(x+3)(x-1) \to \pm\infty$ as x → -3 and as x → 1, and these are
vertical asymptotes. y = 1 is a horizontal asymptote since f(x) =
$(1+3/x+2/x^2)/(1+2/x-3/x^2) \to 1$ as x → ±∞.

33. HINT. $f(x) = (x+4)/(x^2-16) = 1/(x-4)$ if x ≠ -4. Thus x = -4 is not a
vertical asymptote since a finite limit exists as x → -4.

34. $f(x) = \dfrac{\sqrt[3]{(4-x)(4+x)}}{4-x} = \sqrt[3]{\dfrac{(4+x)}{(4-x)^2}} \to \infty$ as x → 4 so

that x = 4 is a vertical asymptote. Dividing numerator and denominator of the
original f(x) formula by x

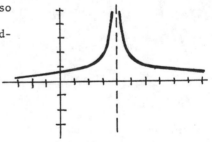

$f(x) = \dfrac{\sqrt[3]{16/x^2 - 1/x}}{4/x - 1} \to \dfrac{0}{-1}$ as x → ±∞.

Thus y = 0 is a horizontal asymptote.

37. $f'(x) = (3/5)x(x+2)^{-2/5} + (x+2)^{3/5}$. As x → -2 the first term becomes in-
finite and the second term approaches 0. Thus $|f'(x)| \to \infty$ as x → -2. Thus
there is a vertical tangent line when x = -2 at the point (-2,0).

40. $f'(x) = (1/3)x^{-2/3} = 1/3x^{2/3} \to \infty$ as x → 0. Thus there is a vertical tangent
when x = 0 at the point (0,-5).

EXERCISES 4.6, page 185

1. Let x and y be the 2 desired numbers with x > y, x-y = 40, and P = xy is to

be minimized. Since $y = x-40$, $P = x(x-40) = x^2 - 40x$ and $P' = 2x-40 = 0$
when $x = 20$. This value minimizes P since $P'' = 2 > 0$. Thus $x = 20$, $y = -20$.

4. Let x be the length of the side of the square base, and y the height. Then
$x^2y = 4$ so that $y = 4/x^2$. If S is the total surface area, then

$$S = 4 \cdot \text{(Area of one side)} + 2 \cdot \text{(Area of base)}$$
$$= 4 \cdot xy \qquad\qquad + 2 \cdot x^2$$
$$= 4x(4/x^2) + 2x^2 = 16/x + 2x^2.$$

Thus $S' = -16/x^2 + 4x = 0$ if $-16 + 4x^3 = 0$ or $x^3 = 4$. Thus $x = \sqrt[3]{4}$ and $y =$
$4/4^{2/3} = \sqrt[3]{4}$ and the box of least surface area (since $S''(\sqrt[3]{4}) > 0$) is a cube
$\sqrt[3]{4}$ on a side.

7. Placing the circle as shown (with equation $x^2 + y^2 = a^2$)
and the rectangle in the upper half, if (x,y) is the point
shown, then the area, A, is $A = 2xy$.

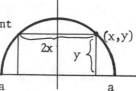

METHOD 1 The Explicit Method. From the equation of
the circle, $y = (a^2-x^2)^{1/2}$ and $A = 2x(a^2-x^2)^{1/2}$, $0 \le x \le a$. $A' =$
$2x(a^2-2x^2)/\sqrt{a^2-x^2} = 0$ when $x = a/\sqrt{2}$ and thus $y = a/\sqrt{2}$ also. This produces a
maximum for A since $A=0$ when $x=0$ or $x=a$ and $A > 0$ otherwise. The dimensions of
the rectangle of maximum area of $x = a/\sqrt{2}$ by $2y = \sqrt{2}a$.

METHOD 2 The Implicit Method. As above, $A = 2xy$ and $A' = 2xy' + 2y$. We obtain y' by differentiating the equation of the circle implicitly: $2x + 2yy'$
$= 0$ or $y' = -x/y$. Substituting into A' we get $A' = -2x^2/y + 2y = 2(y^2-x^2)/y$
$= 0$ when $x = y$. Substituting this into the equation of the circle we get
$2x^2 = a^2$ or $x = y = a/\sqrt{2}$ as in Method 1.

10. Let r and h be the base radius and height of the
cylinder. Then $r^2 = a^2 - h^2/4$. If V is the
volume of the cylinder, then $V = \pi r^2 h =$
$\pi(a^2h - h^3/4)$, for $0 \le h \le a$. $V' = \pi(a^2 - 3h^2/4)$
$= 0$ if $3h^2 = 4a^2$ or $h = 2a/\sqrt{3}$. The corresponding
r is $\sqrt{2/3}a$. This value of h produces a maximum
for V on physical grounds or by an analysis as
in #7 above.

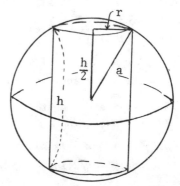

13. The capacity of the trough will be maximized when
the cross-sectional area, A, is maximized. Let
y be the amount turned up and x the base. Then
$x + 2y = 12$ and $A = xy = (12-2y)y$, $0 \le y \le 6$.
$A' = 12-4y = 0$ when $y = 3$, which maximizes A
since $A'' = -4 < 0$.

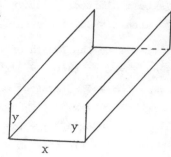

16. Let x and y be the dimensions of the rectangle with
the rotation about an edge of length y. Then $2x +$
$2y = p$ and the volume, V, of the cylinder is $V =$
$\pi x^2 y = \pi x^2 (\frac{p}{2} - x) = \pi(\frac{px^2}{2} - x^3)$. $V' = \pi(px - 3x^2)$
$= 0$ when $x = 0$ or $x = p/3$. The maximum of V occurs
when $x = p/3$ ($V = 0$ when $x = 0$) and the corresponding y is $p/6$. Thus, for maximum volume, the base,
x, should be twice the height y.

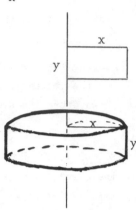

19. If $f(x)$ is the square of the distance of a point (x, x^2+1) on the parabola
from $(3,1)$, then $f(x) = (x-3)^2 + x^4$. $f'(x) = 2(x-3) + 4x^3 = 0$ only when
$x = 1$ and the point is $(1,2)$. Since $f''(1) = 14 > 0$, this is a minimum. (It's
also obvious from a sketch.)

22. Referring to Example 5, the change to be made is that the time on the water
is now $\sqrt{x^2+4}/15$ so that $T = \frac{\sqrt{x^2+4}}{15} + \frac{6-x}{5}$, $0 \le x \le 6$; $\therefore T' = \frac{x}{15\sqrt{x^2+4}} - \frac{1}{5} =$
$\frac{5x - 15\sqrt{x^2+4}}{75\sqrt{x^2+4}}$ which is always negative if $x \ge 0$, $(15\sqrt{x^2+4} > 15x \ge 5x)$. Thus
the minimum of T occurs at the right end point $x = 6$, i.e., he should stay on
the water the entire way.

25. Let x and y be the dimensions shown and A the area. Then
$7x + 2y = 100 \implies y = 1/2(100 - 7x)$. $A = xy = 1/2x(100 -$
$7x) = 50x - \frac{7}{2}x^2 \implies dA/dx = 50 - 7x = 0$ when $x = 50/7$
and $y = 1/2(100 - 7(50/7)) = 25$.

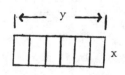

28. Let x be the distance between C and the point opposite A. Then $0 \le x \le 3$
for neither $x < 0$ nor $x > 3$ can produce a minimum cost. Let k be the cost
of pipe per mile above ground. Then 4k is the cost per mile under water.
The above ground length is $3-x$ so that the above ground cost is $k(3-x)$. The
under water length is $\sqrt{1+x^2}$ so that the under water cost is $4k\sqrt{1+x^2}$. If

$C(x)$ is the total cost then $C(x) = k(3-x) + 4k\sqrt{1+x^2}$,

$0 \le x \le 3$. $C'(x) = -k + \dfrac{4kx}{\sqrt{1+x^2}} = 0$ if $\dfrac{4x}{\sqrt{1+x^2}} = 1 \Longrightarrow 4x$

$= \sqrt{1+x^2} \Longrightarrow 16x^2 = 1 + x^2 \Longrightarrow 15x^2 = 1 \Longrightarrow x = \sqrt{1/15}$

miles. That this minimizes $C(x)$ can be checked by computing $C''(x) = 4k/(1+x^2)^{3/2} > 0$, or by evaluating $C(0)$

$= 7k$, $C(3) = 4\sqrt{10}k \approx 12.64$ k, and $C((\sqrt{1/15}) = (\sqrt{15}+3)k$

≈ 6.87 k.

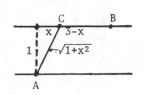

31. Let the point on the ground be x ft. from the 6' pole
and $10-x$ ft. from the 8' pole, and let L be the cable
length. Then $L = \sqrt{x^2+36} + \sqrt{(10-x)^2+64}$, $0 \le x \le 10$.

$\dfrac{dL}{dx} = \dfrac{x}{\sqrt{x^2+36}} + \dfrac{(10-x)(-1)}{\sqrt{(10-x)^2+64}}$

$= \dfrac{x\sqrt{(10-x)^2+64} - (10-x)\sqrt{x^2+36}}{\sqrt{x^2+36}\ \sqrt{(10-x)^2+64}}$. $\dfrac{dL}{dx} = 0$ if the

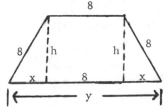

numerator is 0, i.e. when $x\sqrt{(10-x)^2+64} = (10-x)\sqrt{x^2+36}$. Squaring:
$x^2[(10-x)^2+64] = (10-x)^2(x^2+36)$. Cancelling $x^2(10-x)^2$ from each side yields
$64x^2 = (10-x)^2(36) \Longrightarrow 16x^2 - 9(100 - 20x + x^2) = 0 \Longrightarrow 7x^2 + 180x - 900 = 0$.
Using the quadratic formula, or seeing the factorization $(7x-30)(x+30)$, we
obtain $x = 30/7$ as the only critical number in $[0,10]$. L is a minimum there
since $L(0) = 6 + \sqrt{164} \approx 18.8$, $L(10) = \sqrt{136} + 8 \approx 19.7$, whereas $L(30/7) \approx 17.2$.

34. For the fourth side to be parallel to one of the 8'
sides, the trapezoid must be symmetric as shown. With
x as shown, the fourth side has length $y = 8 + 2x$,
$0 \le x \le 8$ (if $x > 8$, then the fourth side is longer
than the sum of the other three sides), and the
height is $h = \sqrt{64-x^2}$. The area, then, is

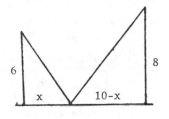

$A = \dfrac{(8+y)}{2} h = (8+x)\sqrt{64-x^2}$.

$\dfrac{dA}{dx} = \dfrac{(8+x)(-x)}{\sqrt{64-x^2}} + \sqrt{64-x^2} = \dfrac{-8x - x^2 + (64-x^2)}{\sqrt{64-x^2}}$

$= -(2x^2 + 8x - 64)/\sqrt{64-x^2} = 0$ if $2x^2 + 8x - 64 = 0 \Longrightarrow x^2 + 4x - 32 =$
$(x-4)(x+8) = 0 \Longrightarrow x = 4$ is the only critical number in $[0,8]$. By evaluating $A(0) = 64$, $A(4) = 12\sqrt{48} \approx 83.1$, $A(8) = 0$, $x = 4$ maximizes A. The
fourth side then has length $y = 8 + 2(4) = 16$.

37. Let x be the side of the square base, g the girth (g = 4x),
     and ℓ the length in inches. Postal regulations require
     ℓ + g ≤ 96. Clearly, to maximize the volume, V, we must
     take ℓ + g = 96, or ℓ + 4x = 96 ⟹ ℓ = 96 - 4x. Then V
     = $x^2ℓ$ = $96x^2 - 4x^3$, 0 < x < 24. dV/dx = $192x - 12x^2$ =
     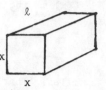
     12x(16-x) = 0 when x = 16. (V"(18) = -192 ⟹ V is maximized.) The correspond-
     ing length is ℓ = 96 - 4(16) = 32.

EXERCISES 4.7, page 193

1.   (a) With r(t) = $3(t+8)^{1/3}$, r'(8) is sought. r'(t) = $(t+8)^{-2/3}$ = $1/\sqrt[3]{(t+8)^2}$
     ⟹ r'(8) = $1/\sqrt[3]{16^2}$ = $1/4\sqrt[3]{4}$.
     (b) Recall that the volume, V, of a sphere of radius r is V = $\frac{4}{3}\pi r^3$. As a
     a function of t, V = $\frac{4}{3}\pi[3\sqrt[3]{t+8}]^3$ = 36π(t+8) ⟹ $D_t V$ = 36π for all t.
     (c) Recall that the surface area, A, of a sphere of radius r is A = $4\pi r^2$
     = $4\pi(3(t+8)^{1/3})^2$ = $36\pi(t+8)^{2/3}$ ⟹ $\frac{dA}{dt}$ = $\frac{2}{3}\cdot 36\pi \cdot (t+8)^{-1/3}$. At t = 8, this
     becomes $\frac{dA}{dt}$ = 24π · $\frac{1}{\sqrt[3]{16}}$ = $6\pi\sqrt[3]{4}$.

     (NOTE: (b) and (c) could also be done using the chain rule, e.g. V = $\frac{4}{3}\pi r^3$
     ⟹ $\frac{dV}{dt}$ = $4\pi r^2 \frac{dr}{dt}$ and substitute the values of r and $\frac{dr}{dt}$ when t = 8.).

4.   By (4.29) we seek T'(t) at t = 2,5,9. T'(t) = $4 - 3/(t+1)^2$.
     (a) T'(2) = 4 - 3/9 = 11/3.   (b) T'(5) = 4 - 3/36 = 47/12.   (c) T'(9) = 4 -
     3/100 = 3.97.

7.   v(t) = s'(t) = 6t-12 = 6(t-2), a(t) = v'(t) = 6, 0 ≤ t ≤ 5. Since v(t) < 0
     for 0 ≤ t < 2 and > 0 for 2 < t ≤ 5, the motion is to the left from s(0) = 1
     to s(2) = -11 from t = 0 to t = 2, then to the right to s(5) = 16 from t = 2
     to t = 5.

10. v(t) = $6-3t^2$ = $3(2-t^2)$, a(t) = -6t, -2 ≤ t ≤ 3. Since v(t) > 0 if |t| < √2
     and < 0 if |t| > √2, we have the following motion: to the left from s(-2) =
     20 to s(-√2) = 24 - 4√2 ≈ 18.4 as t goes from -2 to
     -√2; then to the right until t = √2 and s(√2) = 24
     + 4√2 ≈ 29.66; finally to the left until t = 3 and
     s(3) = 15.

13. v(t) = $8t^3 - 12t$ = $4t(2t^2-3)$, a(t) = $24t^2 - 12$, -2 ≤ t ≤ 2. On the intervals
     [-2,-√(3/2)) and (0,√(3/2)), v(t) < 0 and the motion is to the left. On (-√(3/2),0)
     and (√(3/2),2], v(t) > 0 and the motion is to the right.

16. v(t) = $(1/3)t^{-2/3}$ = $1/3\sqrt[3]{t^2}$, a(t)= $-2/9\sqrt[3]{t^5}$. Since v(t) > 0 for all t, the

motion is strictly to the right from $s(-8) = -2$ to $s(0) = 0$ as t goes from -8 to 0.

19. $v(t) = 6t^2 - 30t + 48$, $a(t) = 12t - 30$. $v(t) = 12 \iff 6t^2 - 30t + 48 = 12$ $\iff 6(t^2 - 5t + 6) = 0 \implies t = 2$ and $t = 3$, $a(2) = 24 - 30 = -6$, $a(3) = 36 - 30 = 6$. For the next part, $a(t) = 10 \iff 12t - 40 = 0 \iff t = 10/3$. $v(10/3) = (600/9) - 100 + 48 = 44/3$.

22. $V = \frac{4}{3}\pi r^3 \implies \frac{dV}{dr} = 4\pi r^2 = A$.

25. We seek $\frac{dF}{dC}$ here. Solving $C = \frac{5}{9}(F-32)$ for F as a function of C we obtain

$F = \frac{9}{5}C + 32 \implies \frac{dF}{dC} = \frac{9}{5}$ for all C.

28. If k is the constant of proportionality, then $R = \frac{k}{d^2}$ and $D_d R = \frac{-2k}{d^3}$. (Don't use the Leibnitz notation if d is the independent variable. $\frac{dR}{dd}$ would result, with 2 meanings for the d's.)

EXERCISES 4.8, page 198

1. Let x and y be the distances shown. Then $x^2 + y^2 = 20^2$; it is given that $dx/dt = 3$, and dy/dt is desired when $y = 8$, (and thus $x = \sqrt{400-64} = \sqrt{336}$). $x^2 + y^2 = 400 \implies$

$2x\frac{dx}{dt} + 2y\frac{dy}{dt} = 0 \implies \frac{dy}{dt} = -\frac{x}{y}\frac{dx}{dt} = -\frac{\sqrt{336}}{8}(3) \approx -6.9$ ft/sec.

(Negative since it's falling and y is decreasing.)

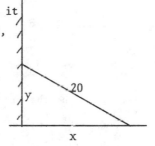

4. Let x be the distance of the first girl east of A, y the distance of the second girl north of A, z the distance between them. Then $z^2 = x^2 + y^2$; $2z\frac{dz}{dt}$ $= 2x\frac{dx}{dt} + 2y\frac{dy}{dt}$. When $t = 2$ min = 120 sec, $x = 1200$ since $\frac{dx}{dt} = 10$. The second girl has only been running for 60 seconds, and so $y = 480$ since $\frac{dy}{dt} = 8$. At that time $z = \sqrt{1200^2 + 480^2} = 120\sqrt{116}$ and substituting these values:

$2(120)\sqrt{116}\,\frac{dz}{dt} = 2(1200)(10) + 2(480)(8) = 240(132) \implies \frac{dz}{dt} = \frac{132}{\sqrt{116}}$

≈ 12.3 ft/sec.

7. Let L = the thickness of the ice so that $\frac{dL}{dt} = -\frac{1}{4}$ in/hr. Let V = volume of the ice so that $\frac{dV}{dt}$ is desired when $L = 2$ in. Now, V=(volume of a hemisphere of radius $(120+L)''$) - (volume of a hemisphere of radius $120''$) =

$\frac{1}{2}(\frac{4}{3})\pi(120+L)^3 - \frac{1}{2}(\frac{4}{3})\pi(120)^3$ $\frac{dV}{dt} = \frac{dV}{dL}\frac{dL}{dt} = 2\pi(120+L)^2\frac{dL}{dt}$, which, when $L = 2$,

yields $\frac{dV}{dt} = 2\pi(122)^2(-\frac{1}{4}) \approx -23{,}368$ in^3/hr.

10. Let y be the altitude of the balloon in feet, and L the distance between it and the observer. Then $dy/dt = 2$ and we want dL/dt when $y = 500$. $L^2 = y^2 + 300^2 \implies$ $2L\, dL/dt = 2y\, dy/dt \implies dL/dt = (y/L)dy/dt$. When $y = 500$, $L = \sqrt{340{,}000} = 100\sqrt{34}$. Thus $dL/dt = 10/\sqrt{34} \approx$ 1.7 ft/sec.

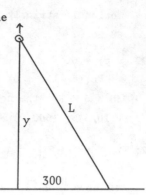

13. Let V be the volume and h the depth of the water. If A is the area of the wetted triangular region at the end of the trough as pictured, then $V = 8A$. The region is an equilateral triangle of side length s and altitude h. Then $s^2 = h^2 + s^2/4 \implies s = 2h/\sqrt{3}$. Also, $A = sh/2 = h^2/\sqrt{3}$ and $V = 8h^2/\sqrt{3}$. Thus $\dfrac{dV}{dt} = \dfrac{16h}{\sqrt{3}}\dfrac{dh}{dt}$. Using $\dfrac{dV}{dt} = 5$, $h = 8'' =$

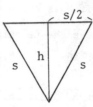

$\dfrac{2'}{3}$, we get $5 = \dfrac{32}{3\sqrt{3}}\dfrac{dh}{dt}$ or $\dfrac{dh}{dt} = \dfrac{15\sqrt{3}}{32} \approx .8$ ft/min ≈ 9.75 in/min.

16. $y^2 = x^2 - 9 \implies 2y\dfrac{dy}{dt} = 2x\dfrac{dx}{dt} = 2x(\dfrac{1}{x}) = 2$. Thus $\dfrac{dy}{dt} = \dfrac{1}{y}$ which at $(5,4)$ is $\dfrac{1}{4}$.

19. Let C be the circumference and r the radius. Then $C = 2\pi r$. We are given that $\dfrac{dr}{dt} = 0.5$ m/sec, and we seek $\dfrac{dC}{dt}$ when $r = 4$. $\dfrac{dC}{dt} = 2\pi\dfrac{dr}{dt} = 2\pi(.5) = \pi$ m/sec for all r.

22. At the given instant, it is given that $p = 40$, $\dfrac{dp}{dt} = 3$, $v = 60$, and we are to find $\dfrac{dv}{dt}$ then. Using the product rule, $pv^{1.4} = c \implies 1.4\, pv^{0.4}\dfrac{dv}{dt} + v^{1.4}\dfrac{dp}{dt}$

$= 0$ so that $\dfrac{dv}{dt} = -\dfrac{v^{1.4}}{1.4pv^{0.4}}\dfrac{dp}{dt} = -\dfrac{v}{1.4p}\dfrac{dp}{dt}$. So: $\dfrac{dv}{dt} = -\dfrac{60}{1.4(40)}(3) = -\dfrac{45}{14} \approx$

-3.2 cm^3/sec., negative since v decreases as p increases (at constant temp.).

25. Let $t = 0$ be the instant at which the first stone is dropped. Then the second is dropped 2 seconds later at $t = 2$. If $2 \le t \le 3$, the first stone has fallen $16t^2$ ft. However, the second stone has been falling for only $t-2$ seconds, and has fallen, therefore, $16(t-2)^2$ ft. The distance, L, between them is then $L = 16t^2 - 16(t-2)^2 = 64t - 64$ and $\dfrac{dL}{dt} = 64$ for all t (i.e. the distance between them steadily increases at the rate of 64 ft/sec.).

28.   Let t = # of hours past 10 a.m.,

      y = # of miles car is north of P,

      x = # of miles plane is <u>west</u> of P,

      z = altitude of plane in miles, and

      L = distance between car and plane.

Then $L^2 = x^2 + y^2 + z^2$. We are given that $\frac{dy}{dt} = 50$,

$\frac{dx}{dt} = -200$ (negative since plane is moving east and x

is decreasing), $\frac{dz}{dt} = 0$ (altitude is constant). We have to find $\frac{dL}{dt}$ when $t = \frac{1}{4}$

(10:15 a.m. is $\frac{1}{4}$ hour past 10). At $t = \frac{1}{4}$, $y = 50(\frac{1}{4}) = 12.5$, $x = 100 - 200(\frac{1}{4})$

$= 50$ (it started 100 miles west of P when t = 0, then moved east for $\frac{1}{4}$ hour),

$z = 5$, and $L = \sqrt{12.5^2 + 50^2 + 5^2} = \sqrt{2681.25} \approx 51.78$. Now, $L^2 = x^2 + y^2 + z^2$

$\implies 2L\frac{dL}{dt} = 2x\frac{dx}{dt} + 2y\frac{dy}{dt} + 2z\frac{dz}{dt} \implies \frac{dL}{dt} = \frac{1}{L}(x\frac{dx}{dt} + y\frac{dy}{dt})$ since $\frac{dz}{dt} = 0$.

Substituting: $\frac{dL}{dt} = \frac{1}{\sqrt{2681.25}}(50(-200) + \frac{25}{2}(50)) = \frac{-9375}{\sqrt{2681.25}} \approx -181$ mph.

EXERCISES 4.9, page 206

NOTE: In Exercises 1-13, F will denote the most general antiderivative of the given function. (4.35)-(4.37) are the formulas used to find F. Remember: The answer F can always be checked by differentiating, F' should be f.

1. $f(x) = 9x^2 - 4x + 3 \implies F(x) = \frac{9x^3}{3} - \frac{4x^2}{2} + 3x + C = 3x^3 - 2x^2 + 3x + C$.

4. $f(x) = 10x^4 - 6x^3 + 5 \implies F(x) = \frac{10x^5}{5} - \frac{6x^4}{4} + 5x + C = 2x^5 - \frac{3}{2}x^4 + 5x + C$.

7. $f(x) = 3\sqrt{x} + 1/\sqrt{x} = 3x^{1/2} + x^{-1/2} \implies F(x) = \frac{3x^{3/2}}{(3/2)} + \frac{x^{1/2}}{(1/2)} + C = 2x^{3/2} +$

    $2x^{1/2} + C$.

10. $f(x) = 3x^5 - \sqrt[3]{x^5} = 3x^5 - x^{5/3} \implies F(x) = \frac{3x^6}{6} - \frac{x^{8/3}}{(8/3)} + C = \frac{x^6}{2} - \frac{3}{8}x^{8/3} + C$.

13. $f(x) = (3x-1)^2 = 9x^2 - 6x + 1 \implies F(x) = \frac{9x^3}{3} - \frac{6x^2}{2} + x + C = 3x^3 - 3x^2 + x + C$.

16. $f(x) = \frac{2x^2-x+3}{\sqrt{x}} = 2x^{3/2} - x^{1/2} + 3x^{-1/2} \implies F(x) = \frac{2x^{5/2}}{(5/2)} - \frac{x^{3/2}}{(3/2)} + \frac{3x^{1/2}}{(1/2)} + C =$

    $\frac{4}{5}x^{5/2} - \frac{2}{3}x^{3/2} + 6x^{1/2} + C$.

19. $f(x) = \frac{x^3-1}{x-1} = x^2 + x + 1$ (if $x \neq 1$) $\implies F(x) = \frac{x^3}{3} + \frac{x^2}{2} + x + C$ (if $x \neq 1$).

22. $f'(x) = 9x^2 + x - 8 \implies f(x) = 3x^3 + \frac{x^2}{2} - 8x + C$. $f(-1) = 1 \implies -3 + (1/2)$

    $+ 8 + C = 1 \implies C = -9/2$. Thus $f(x) = 3x^3 + x^2/2 - 8x - 9/2$.

25. $a(t) = 2-6t \implies v(t) = 2t-3t^2+C.$ $v(0) = -5 \implies C = -5 \implies v(t) = 2t-3t^2-5$
$\implies s(t) = t^2 - t^3 - 5t + D.$ $s(0) = 4 \implies D = 4.$

28. We represent the motion as in Example 5--origin at ground level, positive
direction upward. Then $v(0) = 0$, $s(0) = 1000$. $a(t) = -32 \implies v(t) = -32 + C.$
But $C = 0$ from $v(0) = 0$. So, $v(t) = -32t$ and $s(t) = -16t^2 + 1000$, where s is
the distance above ground. Thus the distance <u>fallen</u> in t seconds is $16t^2$,
and $v(3) = -96$. It strikes the ground when $s(t)$ becomes 0, i.e. when $16t^2$
$= 1000$ or $t = \sqrt{250}/2 \approx 7.9$ sec. (Alternate method: place origin at the 1000
ft. level. Then $s(0) = 0$, $s(t) = -16t^2$, and it hits the ground when $s(t) =$
-1000.)

31. The same setup as in #28 is needed so that $s(t)$ is the distance above ground.
The initial conditions are $v(0) = v_0$, $s(0) = s_0$. $a(t) = -g \implies v(t) = -gt +$
C. $v(0) = v_0 \implies C = v_0 \implies v(t) = -gt + v_0 \implies s(t) = \frac{1}{2}gt^2 + v_0t + D.$
$s(0) = s_0 \implies D = s_0.$

34. Let $a(t) = k$, unknown. The initial and final conditions are $v(0) = 60$ mph $=$
88 ft/sec and $v(9) = 0$. $a(t) = k \implies v(t) = kt + C.$ $v(0) = 88 \implies C = 88$
$\implies v(t) = kt + 88.$ $v(9) = 0 \implies 9k + 88 = 0 \implies k = -88/9$ ft/sec$^2 =$
$-60/9$ mph/sec.

37. $\frac{dV}{dt} = 3\sqrt{t} + \frac{t}{4} = 3t^{1/2} + \frac{t}{4} \implies V = \frac{3t^{3/2}}{(3/2)} + \frac{t^2}{8} + C = 2\sqrt{t^3} + \frac{t^2}{8} + C.$ Setting
$t = 4$ and $V = 20$: $20 = 2\sqrt{64} + \frac{16}{8} + C = 16 + 2 + C \implies C = 2.$

EXERCISES 4.10, page 214

1. (a) $C(100) = 800 + .04(100) + .0002(10,000) = 800 + 4 + 2 = 806.$
(b) Average cost $= c(x) = C(x)/x = 800/x + .04 + .0002x$, $c(100) = 8 + .04 +$
$.02 = 8.06.$ Marginal cost $= C'(x) = .04 + .0004x$, $C'(100) = .04 + .04 = .08.$
(c) $c(x)$ is minimized when $C'(x) = c(x)$ (see p. 209), or $xC'(x) = C(x) \implies$
$.04x + .0004x^2 = 800 + .04x + .0002x^2 \implies .0002x^2 = 800 \implies x^2 = 800/.0002 =$
$4,000,000 \implies x = 2,000.$ The minimum average cost is $c(2,000) = 800/2000 +$
$.04 + .0002(2000) = .40 + .04 + .40 = .84.$
(d) $c'(x) > 0$ for all $x \geq 0$ from (b). Thus $C(x)$ is an increasing function
and the minimum of $C(x)$ is at $x = 0.$

4. (a) $C(100) = 200 + 100/\sqrt{100} + \sqrt{100}/1000 = 200 + 10 + .01 = 210.01.$
(b) $c(x) = C(x)/x = 200/x + 100/x\sqrt{x} + 1/1000\sqrt{x}$, $c(100) = 2 + 1/10 + 1/10,000$
$= 2.1001.$ $C'(x) = -50x^{-3/2} + x^{-1/2}/2000$, $C'(100) = -50/(100)10 + 1/(2000)10$
$= -.05 + .00005 = -.04995.$
(c) Directly, $c'(x) = -200/x^2 = -150x^{-5/2} - x^{-3/2}/2000 < 0$ for all $x > 0$. Thus

c(x) decreases and has no minimum.

(d) $C'(x) = 0 \iff -50x^{-3/2} + x^{-1/2}/2000 = 0 \iff -50 + x/2000 = 0 \iff x = 50(2000) = 100,000.$

7. $3S + 4x - 800 = 0 \implies S = 800/3 - (4/3)x = p(x)$ and $p'(x) = -4/3$, the demand and marginal demand functions. $R(x) = xp(x) = \frac{800}{3}x - \frac{4}{3}x^2$, $R'(x) = \frac{800}{3} - \frac{8}{3}x$, the total revenue and marginal revenue functions. $R'(x) = 0 \implies x = 100$ and $R''(x) < 0 \implies 100$ units for maximum revenue. $p(100) = 800/3 - 400/3 = 400/3 = 133.33.$

10. $Sx^2 - 1000x + 144S - 5000 = 0 \implies S = 1000(x+5)/(x^2+144) = p(x)$ and $p'(x) = 1000(144 - 10x - x^2)/(x^2+144)^2$. $R(x) = xp(x) = 1000(x^2+5x)/(x^2+144)$ and $R'(x) = 1000(720 + 288x - 5x^2)/(x^2+144)^2 = 1000(60-x)(12+5x)/(x^2+144)^2$. $R'(x) = 0$ and $x > 0$ only if $x = 60$. (This maximizes $R(x)$ by the first derivative test.) So, 60 units for maximum revenue. Price per item is $p(60) = 1000(65)/3744 \approx 17.36.$

13. (a) $R(x) = 300x - x^3 \implies R'(x) = 300 - 3x^2 = 3(100-x^2).$
    (b) $R'(x) = 0$ and $x > 0$ only if $x = 10$. $R''(10) = -60 \implies$ maximum total revenue is $R(10) = 3000 - 1000 = 2000.$
    (c) $R(x) = xp(x) \implies p(x) = R(x)/x = 300 - x^2.$

16. $p(x) = 80 - \sqrt{x-1}$, $C(x) = 75x + 2\sqrt{x-1}.$
    (a) $p'(x) = -1/2\sqrt{x-1}.$ (b) $R(x) = xp(x) = 80x - x\sqrt{x-1}.$
    (c) $P(x) = R(x) - C(x) = 5x - (x+2)\sqrt{x-1}.$
    (d) $P'(x) = 5 - \dfrac{(x+2)}{2\sqrt{x-1}} - \sqrt{x-1} = \dfrac{10\sqrt{x-1} - (x+2) - 2(x-1)}{2\sqrt{x-1}} = (10\sqrt{x-1}-3x)/2\sqrt{x-1}.$
    (e) $P'(x) = 0$ when $10\sqrt{x-1} = 3x \iff 100(x-1) = 9x^2 \iff 9x^2 - 100x + 100 = 0 \iff (9x-10)(x-10) = 0 \implies x = 10/9$ and $x = 10$. We calculate $P(1) = 5$, $P(10/9) = 122/27 \approx 4.52$, $P(10) = 14$ and observe that for $x > 10$, $P'(x) < 0$ and $P(x)$ decreases, eventually becoming negative. Thus the maximum profit is 14 when 10 items are produced.
    (f) $C'(x) = 75 + 1/\sqrt{x-1} \implies C'(10) = 75 + 1/3.$

19. Here, $C(x) = 500 + .02x + .001x^2$ and $p(x) = 8$. Thus $R(x) = 8x$ and $P(x) = R(x) - C(x) = 7.98x - 500 - .001x^2$. $P'(x) = 7.98 - .002x = 0$ when $x = 7.98/.002 = 3990$. $P''(x) < 0 \implies$ this produces a maximum of $P(x)$. The maximum profit is $P(3990) = 7.98(3990) - 500 - .001(3990)^2 = 31,840.20 - 500 - 15,920.10 = 15,420.10$ dollars.

22. $p(x) = ax^2 + b \implies R(x) = xp(x) = ax^3 + bx$
    $R'(x) = 3ax^2 + b = 0$ and $x > 0$ only if $x = \sqrt{-b/3a}$ ($b > 0$, $a < 0 \implies -b/3a > 0$)
    $R''(\sqrt{-b/3a}) < 0 \implies$ maximum.

25. $C'(x) = 20 - .015x \Rightarrow C(x) = 20x - .0075x^2 + D.$ $C(1) = 25 = 20 - .0075 + D$ $\Rightarrow D = 5.0075.$ $C(50) = 20(50) - .0075(2500) + 5.0075 = 1000 - 18.75 +$ $5.0075 \approx 986.26.$

28. $R'(x) = 4(x+2)^{-3/2} \Rightarrow R(x) = -8(x+2)^{-1/2} + D.$ The initial condition is $R(0)$ $= 0.$ (No items sold means no revenue.) Thus $-8/\sqrt{2} + D = 0 \Rightarrow D = +8/\sqrt{2} \Rightarrow$ $R(x) = xp(x) = 8(1/\sqrt{2} - 1/\sqrt{x+2}) \Rightarrow p(x) = 8/x(1/\sqrt{2} - 1/\sqrt{x+2}) \Rightarrow p'(x) =$ $\frac{8}{x}(\frac{1}{2(x+2)^{3/2}}) - \frac{8}{x^2}(\frac{1}{\sqrt{2}} - \frac{1}{\sqrt{x+2}}).$

EXERCISES 4.11 (Review), page 216

1. $6x^2 - 2xy + y^3 = 9 \Rightarrow 12x - 2xy' - 2y + 3y^2y' = 0 \Rightarrow y' = (2y-12x)/(3y^2-2x).$ At $(2,-3),$ $y' = (-6-24)/(27-4) = -30/23.$ Thus

     Tangent: $y+3 = (-30/23)(x-2)$
     Normal: $y+3 = (23/30)(x-2).$

4. $x^2 - 2xy + y^2 - 4 = 0 \Rightarrow 2x - 2xy' - 2y + 2yy' = 0 \Rightarrow y' = 1.$ Thus, no horizontal or vertical tangents. (Writing the given equations as $(x-y)^2 - 2^2 = 0,$ the graph is seen to consist of the 2 parallel lines $x-y = \pm 2.$)

7.
$f'(x) = \frac{-2x}{(1+x^2)^2}$ $\begin{cases} > 0 \text{ if } x < 0 \\ \\ < 0 \text{ if } x > 0 \end{cases}$

Thus f is increasing on $(-\infty,0]$, decreasing on $[0,\infty)$, and has a local maximum $f(0) = 1.$

10. (Continuation of #7)
$f''(x) = \frac{2(3x^2-1)}{(1+x^2)^3}$ $\begin{cases} >0 \text{ if } |x| > 1/\sqrt{3} \\ \\ <0 \text{ if } |x| < 1/\sqrt{3} \end{cases}$

Thus the graph is CU on $(-\infty,-1/\sqrt{3})$ and $(1/\sqrt{3},\infty)$, CD on $(-1/\sqrt{3},1/\sqrt{3})$ with points of inflection at $x = \pm 1/\sqrt{3}.$ $f''(0) = -2$ confirms that there is a local maximum at $x = 0.$

13. With the origin at ground level we have initial conditions $v(0) = -30,$ $s(0)$ $= 900.$ Thus, $a(t) = -32 \Rightarrow v(t) = -32t-30 \Rightarrow s(t) = -16t^2 - 30t + 900.$ $v(5) = -160-30 = -190.$ The ground is struck when $s(t) = 0.$ Using the quadratic formula, rejecting the negative root, $t = [30-\sqrt{900+64(900)}]/(-32) = 30(-1+\sqrt{65})/32 \approx 6.6$ sec.

16. $f(x) = 100 \Rightarrow F(x) = 100x + C.$

19. $f''(x) = x^{1/3} - 5 \Rightarrow f'(x) = (3/4)x^{4/3} - 5x + C.$ $f'(1) = 2 \Rightarrow (3/4) - 5 + C$

$= 2 \implies C = 25/4$. Thus $f(x) = \frac{3}{4}(\frac{3}{7})x^{7/3} - \frac{5}{2}x^2 + \frac{25}{4}x + D$. $f(1) = -8 \implies$

$\frac{9}{28} - \frac{5}{2} + \frac{25}{4} + D = -8 \implies D = -\frac{338}{28} = -169/14$.

22. Let the lengths of the pieces be x and 5-x with the x length bent into a cir-
     cle. The radius of this circle is $r = x/2\pi$ and the area is $\pi r^2 = x^2/4\pi$. The
     perimeter of the square being 5-x gives us (5-x)/4 for the length of a side
     and $(5-x)^2/16$ for the area. The sum of these areas, A, is thus A =
     $x^2/4\pi + (5-x)^2/16$, $0 \le x \le 5$. We desire the maximum of A (for part (a)) and
     the minimum of A (for part (b)) on [0,5]. First, we'll find the critical
     numbers, if any, in [0,5]. $A' = \frac{x}{2\pi} - \frac{(5-x)}{8} = x(\frac{1}{2\pi} + \frac{1}{8}) - \frac{5}{8} = x(\frac{8+2\pi}{16\pi}) - \frac{5}{8} = 0$

     if $x = \frac{5}{8}(\frac{16\pi}{8+2\pi}) = \frac{5\pi}{4+\pi} \approx 2.2$. A attains a local minimum there since A'' > 0.
     Since $A(0) = 25/16 \approx 1.56$, and $A(5) = 25/4\pi \approx 2$, A is maximized when x = 5
     (i.e. the entire wire is bent into a circle), and A is minimized when x =
     $5\pi/(4+\pi)$.

25. From the figure, using similar triangles, $\frac{r}{h} = \frac{4}{12}$ or $r = \frac{h}{3}$.

     Thus $V = (1/3)\pi r^2 h = (\pi/27)h^3$ and $\frac{dV}{dt} = \frac{\pi}{9}h^2\frac{dh}{dt}$. We are given

     that $\frac{dV}{dt} = -10$ ft^3/min., and we desire $\frac{dh}{dt}$ when h = 5. Sub-

     stituting: $-10 = \frac{25\pi}{9}\frac{dh}{dt}$ or $\frac{dh}{dt} = -\frac{90}{25\pi} \approx -1.15$ ft/min.

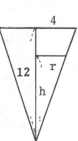

28. $y^2 = 2x^3 \implies 2y\frac{dy}{dt} = 6x^2\frac{dx}{dt}$. At (2,4), $\frac{dy}{dt} = x = 2$ and thus $8(2) = 6(2)^2\frac{dx}{dx}$

     $\implies \frac{dx}{dt} = \frac{16}{24} = \frac{2}{3}$.

31. $\lim\limits_{x\to-\infty} \frac{(2x-5)(3x+1)}{(x+7)(4x-9)} = \lim\limits_{x\to-\infty} \frac{(2-5/x)(3+1/x)}{(1+7/x)(4-9/x)} = \frac{(2)(3)}{(1)(4)} = \frac{3}{2}$.

34. $\lim\limits_{x\to-3} \sqrt[3]{\frac{x+3}{x^3+27}} = \lim\limits_{x\to-3} \sqrt[3]{\frac{x+3}{(x+3)(x^2-3x+9)}} = \lim\limits_{x\to-3} \sqrt[3]{\frac{1}{x^2-3x+9}} = \sqrt[3]{\frac{1}{9+9+9}} = \sqrt[3]{\frac{1}{27}} = \frac{1}{3}$.

37. As $x \to 0^+$, $\sqrt{x} \to 0$ and is positive. Thus $1/\sqrt{x} \to \infty$ and $\sqrt{x} - 1/\sqrt{x} \to -\infty$.

40. $\lim\limits_{x\to+\infty} \frac{x^2}{(1-x)^2} = \lim\limits_{x\to+\infty} \frac{1}{(1-1/x)^2} = 1 \implies$ y = 1 is a horizontal asymptote.

     $\lim\limits_{x\to 1} \frac{x^2}{(1-x)^2} = \infty \implies$ x = 1 is a vertical asymptote. For the graph, f'(x) =

     $-2x/(1-x)^3 = 0$ only at x = 0. $f''(x) = 2(2x+1)/(x-1)^4$. $f''(0) = 2 > 0 \implies$

f(0) = 0 is a local minimum. There is a PI at x = -1/2.

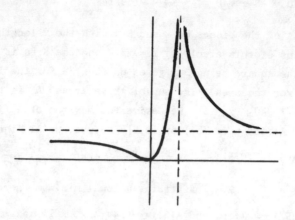

EXERCISES 5.1, page 227

1. $\displaystyle\sum_{k=1}^{5}$ (3k-10) = (3·1-10) + (3·2-10) + (3·3-10) + (3·4-10) + (3·5-10) =

   -7-4-1+2+5 = -5.

4. Observe that $[1+(-1)^k]$ = 2 if k is even and = 0 if k is odd. Thus $\displaystyle\sum_{k=1}^{10}$ $[1+(-1)^k]$

   = 0+2+0+2+0+2+0+2+0+2 = 10.

7. $\displaystyle\sum_{i=1}^{8}$ 2^i = 2^1 + 2^2 + 2^3 + 2^4 + 2^5 + 2^6 + 2^7 + 2^8 = 2+4+8+16+32+64+128+256 = 510.

10. $\displaystyle\sum_{k=1}^{1000}$ 2 = 2+2+...+2, 1000 times yielding 2(1000)= 2000.

13. $\displaystyle\sum_{k=1}^{n}$ (k^2+3k+5) = $\displaystyle\sum_{k=1}^{n}$ k^2 + 3 $\displaystyle\sum_{k=1}^{n}$ k + $\displaystyle\sum_{k=1}^{n}$ 5 = $\dfrac{n(n+1)(2n+1)}{6}$ + $\dfrac{3n(n+1)}{2}$ + 5n

    = $\dfrac{n}{6}$[(n+1)(2n+1) + 9(n+1) + 30] = $n(n^2+6n+20)/3$.

16. $\displaystyle\sum_{k=1}^{n}$ (k^3+2k^2-k+4) = $\displaystyle\sum_{k=1}^{n}$ k^3 + 2 $\displaystyle\sum_{k=1}^{n}$ k^2 - $\displaystyle\sum_{k=1}^{n}$ k + $\displaystyle\sum_{k=1}^{n}$ 4 = $(\dfrac{n(n+1)}{2})^2$ +

    $\dfrac{2n(n+1)(2n+1)}{6}$ - $\dfrac{n(n+1)}{2}$ + 4n = $n[3n(n+1)^2$ + 4(n+1)(2n+1) - 6(n+1) + 48]/12 =

    $n(3n^3 + 14n^2 + 9n + 46)/12$.

19. a = 0, b = 5 $\Longrightarrow$ Δx = (b-a)/n = 5/n, and the subdividing points are x_o = 0,
    x_1 = a + Δx = 5/n, x_2 = a + 2Δx = 10/n, ..., x_i = a + iΔx = 5i/n, ..., x_n = 5.
    Since $f(x)$ = x^2 is increasing on [0,5], the minimum value of f on $[x_{i-1},x_i]$
    occurs at the left end point, x_{i-1}, and the maximum value of f occurs at the
    right end point, x_1. Thus u_i = x_{i-1} = 5(i-1)/n and v_i = x_i = 5i/n.

    (a) Here we need $f(u_i)$ = $f(x_{i-1})$ = $f(\dfrac{5(i-1)}{n})$ =

    $[\dfrac{5(i-1)}{n}]^2$ = $\dfrac{25(i-1)^2}{n^2}$. Then

    $\displaystyle\sum_{i=1}^{n}$ $f(u_i)\Delta x$ = $\displaystyle\sum_{i=1}^{n}$ $\dfrac{25(i-1)^2}{n^2}$ · $\dfrac{5}{n}$ = $\dfrac{125}{n^3}$ $\displaystyle\sum_{i=1}^{n}$ $(i-1)^2$.

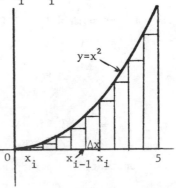

To evaluate the sum we write $\sum\limits_{i=1}^{n} (i-1)^2 =$

$$\sum_{i=1}^{n} (i^2-2i+1) = \sum_{i=1}^{n} i^2 - 2 \sum_{i=1}^{n} i + \sum_{i=1}^{n} 1 =$$

$[\frac{n(n+1)(2n+1)}{6} - 2(\frac{n(n+1)}{2}) + n]$. We used (5.10) to evaluate the first sum,

(5.9) for the second sum, and (5.1) (with c = 1) for the third. Substituting

the value for the sum into the previous expression, we obtain $\sum\limits_{i=1}^{n} f(u_i)\Delta x =$

$\frac{125}{n^3}[\frac{n(n+1)(2n+1)}{6} - n(n+1) + n] = \frac{125(n+1)(2n+1)}{6n^2} - \frac{125(n+1)}{n^2} + \frac{125}{n^2} =$

$\frac{125}{6}(1 + \frac{1}{n})(2 + \frac{1}{n}) - 125(\frac{1}{n} + \frac{1}{n^2}) + \frac{125}{n^2}$. As $\Delta x \to 0$, $\frac{5}{n} \to 0$, $n \to \infty$. Also $\frac{1}{n} \to 0$

and $\frac{1}{n^2} \to 0$ so that $(1 + \frac{1}{n}) \to 1$ and $(2 + \frac{1}{n}) \to 2$. Moreover, in the limit as

$n \to \infty$, the sums, $\sum\limits_{i=1}^{n} f(u_i)\Delta x$, approach the area. Thus, $\sum\limits_{i=1}^{n} f(u_i)\Delta x =$

$\frac{125}{6}(1+\frac{1}{n})(2+\frac{1}{n}) - 125(\frac{1}{n} + \frac{1}{n^2}) + \frac{125}{n^2} \to \frac{125}{6}(1)(2) - 125(0+0) - 125(0) = \frac{125}{3} = A.$

(b) Here we need $f(v_i) = f(x_i) = f(5i/n) = (5i/n)^2$

$= 25i^2/n^2$. Then $\sum\limits_{i=1}^{n} f(v_i)\Delta x = \sum\limits_{i=1}^{n} \frac{25i^2}{n^2} \cdot \frac{5}{n} =$

$\frac{125}{n^3} \sum\limits_{i=1}^{n} i^2 = \frac{125}{n^3} \cdot \frac{n(n+1)(2n+1)}{6} = \frac{125}{6}(1 + \frac{1}{n})$

$(2 + \frac{1}{n})$, where we used (5.10) to evaluate the sum.

As above, these sums approach the area as $n \to \infty$

and we obtain $A = \frac{125}{6}(1)(2) = \frac{125}{3}$.

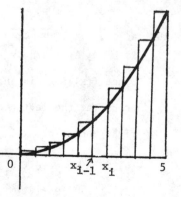

22. a = -2, b = 6 $\Longrightarrow$ $\Delta x = (6-(-2))/n = 8/n$, and the subdividing points are $x_o = a$

   $= -2$, $x_1 = -2 + 8/n$, ..., $x_i = a + i\Delta x = -2 + 8i/n$, ..., $x_n = 6$. Since f is

   constant, any point in $[x_{i-1}, x_i]$ can be used for u_i or v_i. $\sum\limits_{i=1}^{n} f(u_i)\Delta x =$

$$\sum_{i=1}^{n} f(v_i)\Delta x = \sum_{i=1}^{n} 7 \cdot \frac{8}{n} = \frac{56}{n} \sum_{i=1}^{n} 1 = \frac{56}{n}(n) = 56 \text{ for all } \Delta x \text{ and } n. \quad A = \lim_{n \to \infty} 56$$

$= 56$ as expected since the region is a rectangle with base $6-(-2) = 8$ and height 7.

25. Here $a = 1$, $b = 2$, $\Delta x = 1/n$, and the subdividing points are $x_0 = 1$, $x_1 = 1 + 1/n$, ..., $x_i = 1 + i/n$, ..., $x_n = 2$. Since $f(x) = x^3 + 1$ is increasing, the minimum value of f on $[x_{i-1}, x_i]$ occurs at the left end point, x_{i-1}, and the maximum value of f occurs at the right end point, x_i. Thus $u_i = x_{i-1} = 1 + (i-1)/n$ and $v_i = x_i = 1 + i/n$.

(a) Here we need $f(u_i) = f(x_{i-1}) = f(1 + \frac{i-1}{n}) = (1 + \frac{i-1}{n})^3 + 1$. Recalling that $(c+d)^3 = c^3 + 3c^2 d + 3cd^2 + d^3$ we obtain $f(u_i) =$

$[1 + \frac{3(i-1)}{n} + \frac{3(i-1)^2}{n^2} + \frac{(i-1)^3}{n^3}] + 1$. Then $\sum_{i=1}^{n} f(u_i)\Delta x = \sum_{i=1}^{n} [2 + \frac{3(i-1)}{n}$

$+ \frac{3(i-1)^2}{n^2} + \frac{(i-1)^3}{n^3}] \cdot \frac{1}{n} = \frac{1}{n} \sum_{i=1}^{n} 2 + \frac{1}{n} \sum_{i=1}^{n} [\frac{3(i-1)}{n} + \frac{3(i-1)^2}{n^2} + \frac{(i-1)^3}{n^3}]$.

We must now evaluate these sums. We could evaluate the second sum by expanding all the $(i-1)$ expressions as in No. 19(a) above; however, there is an easier way. Observe that $\sum_{i=1}^{n} (i-1)^m = 0^m + 1^m + 2^m + \ldots + (n-1)^m = \sum_{j=1}^{n-1} j^m$ for any positive number m. Thus, (replacing the first sum by 2n using (5.1)),

$\sum_{i=1}^{n} f(u_i)\Delta x = (\frac{1}{n})2n + \frac{1}{n} \sum_{i=1}^{n} [\frac{3(i-1)}{n} + \frac{3(i-1)^2}{n^2} + \frac{(i-1)^3}{n^3}] = 2 + \frac{1}{n} \sum_{j=1}^{n-1} (\frac{3j}{n}$

$+ \frac{3j^2}{n^2} + \frac{j^3}{n^3}) = 2 + \frac{3}{n^2} \sum_{j=1}^{n-1} j + \frac{3}{n^3} \sum_{j=1}^{n-1} j^2 + \frac{1}{n^4} \sum_{j=1}^{n-1} j^3$. Now we use (5.9),

(5.10), (5.11) with n replaced by n-1 (since the upper limit of summation is now n-1) to obtain $\sum_{i=1}^{n} f(u_i)\Delta x = 2 + \frac{3}{n^2} \frac{(n-1)n}{2} + \frac{3}{n^3} \frac{(n-1)n(2n-1)}{6} +$

$\frac{1}{n^4}(\frac{(n-1)n}{2})^2 = 2 + \frac{3}{2} \frac{n-1}{n} + \frac{3}{6} \frac{n-1}{n} \frac{2n-1}{n} + \frac{1}{4}(\frac{n-1}{n})^2$. As $\Delta x \to 0$, $\frac{1}{n} \to 0$, and $n \to \infty$, also $\frac{n-1}{n} = 1 - \frac{1}{n} \to 1$ and $\frac{2n-1}{n} = 2 - \frac{1}{n} \to 2$ and the sums $\sum_{i=1}^{n} f(u_i)\Delta x$ approach the area, A, as n approaches ∞. Thus, letting n approach ∞ in the above ex-

pression for $\sum\limits_{i=1}^{n} f(u_i)\Delta x$, we obtain $A = \lim\limits_{n\to\infty} \sum\limits_{i=1}^{n} f(u_i)\Delta x = \lim\limits_{n\to\infty} [2 + \frac{3}{2}\frac{n-1}{n} +$

$\frac{1}{2}\frac{n-1}{n}\frac{2n-1}{n} + \frac{1}{4}(\frac{n-1}{n})^2] = 2 + \frac{3}{2}(1) + \frac{1}{2}(1)(2) + \frac{1}{4}(1)^2 = 2 + \frac{3}{2} + 1 + \frac{1}{4} =$

$\frac{8+6+4+1}{4} = \frac{19}{4}$.

25. Here $a = 1$, $b = 2$, $\Delta x = 1/n$, and the subdividing points are $x_o = 1$, $x_1 = 1 +$

(b) $f(v_i) = f(x_i) = x_i^3 + 1 = (1 + \frac{i}{n})^3 + 1 = 2 + \frac{3}{n}i + \frac{3}{n^2}i^2 + \frac{i^3}{n^3}$. Then

recalling $\Delta x = 1/n$, $\sum\limits_{i=1}^{n} f(v_i)\Delta x = \sum\limits_{i=1}^{n} (2 + \frac{3}{n}i + \frac{3}{n^2}i^2 + \frac{i^3}{n^3}) \cdot \frac{1}{n} =$

$\frac{1}{n}\sum\limits_{i=1}^{n} 2 + \frac{3}{n^2}\sum\limits_{i=1}^{n} i + \frac{3}{n^3}\sum\limits_{i=1}^{n} i^2 + \frac{1}{n^4}\sum\limits_{i=1}^{n} i^3 = \frac{1}{n}(2n) + \frac{3}{n^2} \cdot \frac{n(n+1)}{2} + \frac{3}{n^3} \cdot$

$\frac{n(n+1)(2n+1)}{6} + \frac{1}{n^4}(\frac{n(n+1)}{2})^2 = 2 + \frac{3}{2}\frac{n+1}{n} + \frac{1}{2}\frac{(n+1)(2n+1)}{n^2} + \frac{1}{4}\frac{(n+1)^2}{n^2}$. As

$\Delta x \to 0$, $n \to \infty$, and, as above, $A = 2 + \frac{3}{2}(1) + \frac{1}{2}(1)(2) + \frac{1}{4}(1)^2 =$

$\frac{8+6+4+1}{4} = \frac{19}{4}$.

EXERCISES 5.2, page 234

1. $x_o = 0$, $x_1 = 1.1$, $x_2 = 2.6$, $x_3 = 3.7$, $x_4 = 4.1$ and $x_5 = 5$. Thus $\Delta x_1 = x_1 - x_0$
   $= 1.1$, $\Delta x_2 = x_2 - x_1 = 1.5$, $\Delta x_3 = x_3 - x_2 = 1.1$, $\Delta x_4 = x_4 - x_3 = 0.4$ and $\Delta x_5 =$
   $x_5 - x_4 = 0.9$. The largest of the Δx_i's is $\Delta x_2 = 1.5 \implies \|P\| = 1.5$.

4. $x_o = 1$, $x_1 = 1.6$, $x_2 = 2$, $x_3 = 3.5$, $x_4 = 4$. $\Delta x_1 = x_1 - x_o = 0.6$, $\Delta x_2 = x_2 - x_1$
   $= 0.4$, $\Delta x_3 = x_3 - x_2 = 1.5$, $\Delta x_4 = x_4 - x_3 = 0.5$. $\|P\| = 1.5$ since the largest
   Δx_i is $\Delta x_3 = 1.5$.

7. Since this is a regular partition, all Δx_i's are equal to $(b-a)/n = (6-0)/6$
   $= 1$. With w_i as the midpoint of $[x_{i-1}, x_i] = [i-1, i]$, we have $R_p =$

   $\sum\limits_{i=1}^{6} f(w_i)\Delta x_i = f(\frac{1}{2}) + f(\frac{3}{2}) + f(\frac{5}{2}) + f(\frac{7}{2}) + f(\frac{9}{2}) + f(\frac{11}{2}) = (8 - \frac{1}{2}(\frac{1}{4})) +$

$$(8 - \frac{1}{2}(\frac{9}{4})) + (8 - \frac{1}{2}(\frac{25}{4})) + (8 - \frac{1}{2}(\frac{49}{4})) + (8 - \frac{1}{2}(\frac{81}{4})) + (8 - \frac{1}{2}(\frac{121}{4})) =$$

$$48 - \frac{1}{2} \frac{(1+9+25+49+81+121)}{4} = 48 - \frac{286}{8} = \frac{49}{4} .$$

10. Here $\Delta x_1 = \Delta x_2 = \Delta x_3 = \Delta x_4 = 2$, $\Delta x_5 = 7$, and $R_p = f(1) \cdot 2 + f(4) \cdot 2 + f(5) \cdot 2 +$

$f(9) \cdot 2 + f(9) \cdot 7 = 1 \cdot 2 + \sqrt{4} \cdot 2 + \sqrt{5} \cdot 2 + \sqrt{9} \cdot 2 + \sqrt{9} \cdot 7 = 2 + 4 + 2\sqrt{5} + 6 + 21 =$

$33 + 2\sqrt{5} \approx 37.47$.

13. According to (5.16), the expression involving w_i in the sum is $f(w_i)$. Here,

$f(w_i) = 2\pi w_i(1+w_i^3)$ so that $f(x) = 2\pi x(1+x^3)$. The interval being partitioned

is $[0,4]$. Thus, by (5.16) the limit is $\int_0^4 2\pi x(1+x^3)dx$.

16. By (5.18), the value of the integral is 0.

19. $f(x) = 2x+6$ is continuous and ≥ 0 on $[-3,2]$. Thus, by (5.20) $\int_{-3}^2 2x+6$ dx is

the area under the graph of f from -3 to 2. The graph of $f(x) = 2x+6$ is a

line from $(-3,0)$ to $(2,10)$, and the region is a triangle with base $2-(-3) =$

5 and height 10. Thus the integral = area = $(1/2)5(10) = 25$.

22. $f(x) = \sqrt{a^2-x^2}$ is continuous and ≥ 0 on $[-a,a]$. By (5.20) the integral is the

area under the graph of f from -a to a. The graph of f is the upper half of

the circle of radius a centered at the origin. $(y = \sqrt{a^2-x^2} \Longrightarrow y \geq 0$ and

$y^2 = a^2 - x^2$, or $x^2 + y^2 = a^2$.) The region is, thus, a semi-circle of radius

a and area $(1/2)\pi a^2$, which is the value of the integral.

EXERCISES 5.3, page 240

1. $\int_{-2}^4 5$ dx $= 5(4-(-2)) = 5(6) = 30$ by (5.21).

4. $\int_4^{-3}$ dx $= -\int_{-3}^4$ dx $= -1(4-(-3)) = -7$ by (5.17) and (5.21).

7. By (5.23) $\int_1^4 (3x^2+5)dx = \int_1^4 3x^2 dx + \int_1^4 5$ dx. By (5.22) and (5.21) we get:

$3\int_1^4 x^2 dx + 5(4-1) = 3(21) + 5(3) = 63 + 15 = 78$.

10. Using (5.23), (5.22) and (5.21), $\int_1^4 (3x+2)^2$ dx $= \int_1^4 (9x^2+12x+4)dx = \int_1^4 9x^2 dx$

$+ \int_1^4 12x$ dx $+ \int_1^4 4$ dx $= 9\int_1^4 x^2 dx + 12\int_1^4 x$ dx $+ 4(4-1) = 9(21) + 12(15/2) + 12$

$= 291$.

13. $\int_{1}^{4} (\sqrt{x}-5)^2 dx = \int_{1}^{4} (x-10\sqrt{x}+25) dx = \int_{1}^{4} x\, dx - 10\int_{1}^{4} \sqrt{x}\, dx + \int_{1}^{4} 25\, dx = \frac{15}{2} -$

$10(\frac{14}{3}) + 25(3) = \frac{45}{6} - \frac{280}{6} + \frac{450}{6} = \frac{215}{6}$.

16. We can show that $f(x) = 5x^2 - 4\sqrt{x} + 2$ is positive for all x in [2,4]. For such $x, 5x^2 + 2 \geq 22$ and $-4\sqrt{x} \geq -4\sqrt{4} = -8$. Adding these inequalities yields $5x^2 + 2 - 4\sqrt{x} \geq 22 - 8$, or $f(x) \geq 14$ on [2,4]. By (5.28),

$\int_{2}^{4} f(x)\, dx \geq \int_{2}^{4} 14\, dx = 14(4-2) = 28 > 0$, as desired.

19. Rewriting the sum as $\int_{-3}^{5} f(x)\, dx + \int_{5}^{1} f(x)\, dx$, we use (5.26) to combine these

to get $\int_{-3}^{1} f(x)\, dx$.

22. $\int_{-2}^{6} f(x)\, dx - \int_{-2}^{2} f(x)\, dx = \left[\int_{-2}^{2} f(x)\, dx + \int_{2}^{6} f(x)\, dx \right] - \int_{-2}^{2} f(x)\, dx = \int_{2}^{6} f(x)\, dx,$

where (5.26) was used to write the first integral as the sum in the brackets.

25. We write $\int_{a}^{b}$ as an abbreviation of $\int_{a}^{b} f(x)\, dx$. The first equation in each of

the four other cases follows from (5.25) for different a, b, c, of course.

(1) $a < b < c.$ $\quad \int_{a}^{c} = \int_{a}^{b} + \int_{b}^{c} \implies \int_{a}^{b} = \int_{a}^{c} - \int_{b}^{c} \implies \int_{a}^{b} = \int_{a}^{c} + \int_{c}^{b}$

(2) $c < b < a.$ $\quad \int_{c}^{a} = \int_{c}^{b} + \int_{b}^{a} \implies -\int_{a}^{c} = \int_{c}^{b} - \int_{a}^{b} \implies \int_{a}^{b} = \int_{a}^{c} + \int_{c}^{b}$

(3) $b < a < c.$ $\quad \int_{b}^{c} = \int_{b}^{a} + \int_{a}^{c} \implies -\int_{c}^{b} = -\int_{a}^{b} + \int_{a}^{c} \implies \int_{a}^{b} = \int_{a}^{c} + \int_{c}^{b}$

(4) $b < c < a.$ $\quad \int_{b}^{a} = \int_{b}^{c} + \int_{c}^{a} \implies -\int_{a}^{b} = -\int_{c}^{b} - \int_{a}^{c} \implies \int_{a}^{b} = \int_{a}^{c} + \int_{c}^{b}$

EXERCISES 5.4, page 244

1. We seek $z \in (0,3)$ so that $(3z^2)(3-0) = 27$ or $z^2 = 3$. Thus $z = \sqrt{3}$.

4. We seek $z \in (-2,-1)$ so that $8z^{-3}(-1-(-2)) = -3$ or $8z^{-3} = -3$. Thus $z = \sqrt[3]{-8/3} \approx -1.39$.

7. We seek $z \in (2,7)$ so that $\frac{1}{(z+3)^2} (7-2) = \frac{1}{10}$ or $\frac{1}{(z+3)^2} = \frac{1}{50}$. Thus $(z+3)^2 = 50$ $\implies z+3 = \pm\sqrt{50} = \pm5\sqrt{2} \implies z = -3 \pm 5\sqrt{2}$. For z to be in (2,7) the "+" sign must be chosen, and $z = -3 + 5\sqrt{2} \approx 4.07$.

10. We seek $z \in (1,3)$ so that $(z^2 + \frac{1}{z^2})(3-1) = \frac{28}{3}$ or $z^2 + \frac{1}{z^2} = \frac{14}{3}$. To solve,

let $y = z^2$, and we obtain $y + \frac{1}{y} = \frac{14}{3}$ or $3y^2 - 14y + 3 = 0 \implies y =$

$\frac{14 \pm \sqrt{14^2 - 4(3)(3)}}{6} = \frac{14 \pm \sqrt{196-36}}{6} = \frac{14 \pm \sqrt{160}}{6} = \frac{7 \pm \sqrt{40}}{3}$. If the "$-$"

sign is used, then $y = z^2 = \frac{7 - \sqrt{40}}{3} \approx \frac{0.52}{3}$ and z is not in $(1,3)$. So we use

the "$+$" sign and $y = z^2 = \frac{(7 + \sqrt{40})}{3} \implies z = \sqrt{(7 + \sqrt{40}/3} \approx 2.1$.

12. $\int_a^b x \, dx$ = area of the trapezoid shown = $\frac{(b+a)}{2}(b-a)$.

Thus we seek $z \in (a,b)$ so that $z(b-a) = \frac{(b+a)}{2}(b-a)$.

Thus $z = \frac{b+a}{2}$.

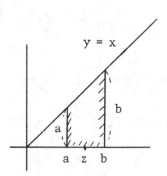

EXERCISES 5.5, page 250

1. $\int_1^4 (x^2-4x-3) \, dx = \frac{x^3}{3} - 2x^2 - 3x \Big]_1^4 = [\frac{64}{3} - 32 - 12] - [\frac{1}{3} - 2 - 3] = -18$.

4. $\int_0^2 (w^4-2w^3) \, dw = \frac{w^5}{5} - \frac{w^4}{2} \Big]_0^2 = [\frac{32}{5} - \frac{16}{2}] - [0-0] = -8/5$.

7. $\int_1^2 \frac{5}{8x^6} \, dx = \frac{5}{8} \int_1^2 x^{-6} \, dx = \frac{5}{8}[\frac{x^{-5}}{-5}]_1^2 = -\frac{1}{8}[2^{-5} - 1^{-5}] = -\frac{1}{8}[\frac{1}{32} - 1] = \frac{31}{256}$.

10. $\int_{-1}^{-2} \frac{2s-7}{s^3} \, ds = \int_{-1}^{-2} (2s^{-2} - 7s^{-3}) \, ds = -2s^{-1} + \frac{7}{2}s^{-2}]_{-1}^{-2} = -\frac{2}{s} + \frac{7}{2s^2}]_{-1}^{-2} =$

$[\frac{-2}{-2} + \frac{7}{2(4)}] - [\frac{-2}{-1} + \frac{7}{2}] = 1 + \frac{7}{8} - 2 - \frac{7}{2} = -\frac{29}{8}$.

13. $\int_0^1 (2x-3)(5x+1) \, dx = \int_0^1 (10x^2-13x-3) \, dx = \frac{10}{3}x^3 - \frac{13}{2}x^2 - 3x]_0^1 = \frac{10}{3} - \frac{13}{2} - 3$

$= -\frac{37}{6}$.

16. The value is 0 since the upper and lower limits are the same. See (5.18).

17. HINT: $(x^2-1)/(x-1) = x+1$ if $x \neq 1$.

19. See #16.

22. $\int_{-2}^{-1} (r - \frac{1}{r})^2 dr = \int_{-2}^{-1} (r^2 - 2r(\frac{1}{r}) + \frac{1}{r^2}) dr = \int_{-2}^{-1} (r^2 - 2 + r^{-2}) dr =$

$\frac{r^3}{3} - 2r - r^{-1}]_{-2}^{-1} = [\frac{(-1)^3}{3} - 2(-1) - \frac{1}{(-1)}] - [\frac{(-2)^3}{3} - 2(-2) - \frac{1}{(-2)}] =$

$[-\frac{1}{3} + 2 + 1] - [-\frac{8}{3} + 4 + \frac{1}{2}] = \frac{7}{3} - \frac{3}{2} = \frac{5}{6}$.

25. $\int_0^4 \sqrt{3t} (\sqrt{t} + \sqrt{3}) dt = \sqrt{3} \int_0^4 \sqrt{t}(\sqrt{t} + \sqrt{3}) dt = \sqrt{3} \int_0^4 (t + \sqrt{3} \, t^{1/2}) dt =$

$\sqrt{3}[\frac{t^2}{2} + \frac{\sqrt{3}}{3/2} t^{3/2}]_0^4 = \sqrt{3}\{[\frac{16}{2} + \frac{2\sqrt{3}}{3} \cdot 4^{3/2}] - [0-0]\} = \sqrt{3}(8 + \frac{2}{\sqrt{3}}(8)) = 8(\sqrt{3} + 2)$.

28. $D_x \int_0^x (5t+3)^2 dt = D_x \int_0^x (25t^2 + 30t + 9) dt = D_x[\frac{25}{3}t^3 + 15t^2 + 9t]_0^x =$

$D_x[\frac{25}{3}x^3 + 15x^2 + 9x] = 25x^2 + 30x + 9 = (5x+3)^2$.

31. Since $f(x) = x^2+1 > 0$, the area $= \int_{-1}^2 (x^2+1) dx = [\frac{x^3}{3} + x]_{-1}^2 = (\frac{8}{3} + 2) -$

$(-\frac{1}{3} - 1) = 6$.

34. Verify by differentiating the function on the right to obtain the integrand.

37. $\int_0^4 (\sqrt{x}+1) dx = \int_0^4 (x^{1/2}+1) dx = \frac{2}{3}x^{3/2} + x]_0^4 = \frac{2}{3}(8) + 4 = \frac{28}{3}$. Thus we seek

$z \in (0,4)$ such that $(\sqrt{z}+1)4 = \frac{28}{3}$ or $\sqrt{z} + 1 = \frac{7}{3} \Longrightarrow \sqrt{z} = \frac{4}{3} \Longrightarrow z = \frac{16}{9}$.

40. $\int_1^9 3x^{-2} dx = -3x^{-1}]_1^9 = -3(\frac{1}{9} - 1) = -3(-\frac{8}{9}) = \frac{8}{3}$. $\frac{3}{z^2}(9-1) = \frac{8}{3} \Longleftrightarrow 8z^2 = 72 \Longleftrightarrow$

$z^2 = 9 \Longrightarrow z = 3$. ($z = -3$ is not in $(1,9)$.)

EXERCISES 5.6, page 257

1. Let $u = 3x+1$ so that $du = 3 \, dx$ or $dx = (1/3) du$. $\int (3x+1)^4 \, dx = (1/3) \int u^4 \, du =$

$u^5/15 + C = \frac{(3x+1)^5}{15} + C$.

4. Using the first method of Example 1, let $u = 9 - z^2$, so $du = -2z \, dz$. Then

$\int \sqrt{9-z^2} \, z \, dz = (-1/2) \int (9-z^2)^{1/2}(-2zdz) = -\frac{1}{2} \int u^{1/2} du = -\frac{1}{2}(\frac{2}{3}) u^{3/2} + C$

$= -\frac{1}{3}(9-z^2)^{3/2} + C$.

7. Let $u = 1-2s^2$, $du = -4s \, ds$, $s \, ds = -\frac{1}{4} du$. $\int \frac{s}{\sqrt[3]{1-2s^2}} \, ds = -\frac{1}{4} \int u^{-1/3} du =$

$-\frac{1}{4}(\frac{3}{2}) u^{2/3} + C = -\frac{3}{8}(1-2s^2)^{2/3} + C$.

10. Let $v = 1 + 1/u$, $dv = -1/u^2 \, du$. Then $\int (1 + \frac{1}{u})^{-3} (\frac{1}{u^2}) du = -\int v^{-3} \, dv = \frac{v^{-2}}{2} + C$

$= \frac{1}{2}(1 + \frac{1}{u})^{-2} + C$.

13. Let $u = t^2-1$, $du = 2t \, dt$. When $t = 1$ or -1, $u = 0$. Thus $\int_{-1}^{1} (t^2-1)^3 \, t \, dt =$

$\frac{1}{2} \int_{0}^{0} u^3 \, du = 0$.

16. Let $u = x^2+9$. Then $du = 2x \, dx$, $x \, dx = (1/2) du$. $x = 0 \Rightarrow u = 9$. $x = 4 \Rightarrow$

$u = 25$. $\int_{0}^{4} \frac{x}{\sqrt{x^2+9}} dx = \int_{0}^{4} (x^2+9)^{-1/2} x \, dx = \int_{9}^{25} u^{-1/2} (1/2) du = u^{1/2}]_{9}^{25} =$

$\sqrt{25} - \sqrt{9} = 5 - 3 = 2$.

19. Let $u = 8z+5$, $du = 8 \, dz$ so that $dz = \frac{1}{8} du$. Then $\int 5(8z+5)^{1/2} dz = \frac{5}{8} \int u^{1/2} \, du$

$= \frac{5}{8} \frac{u^{3/2}}{(3/2)} + C = \frac{5}{8} \cdot \frac{2}{3} (8z+5)^{3/2} + C$.

22. Let $u = 3-x^4$, $du = -4x^3 dx$ so that $x^3 dx = -\frac{1}{4} du$. Then $\int (3-x^4)^3 \, x^3 dx =$

$-\frac{1}{4} \int u^3 du = -\frac{1}{4} \frac{u^4}{4} + C = -\frac{1}{16} (3-x^4)^4 + C$.

25. Let I denote the given integral.

(a) Let $u = \sqrt{x} + 3$. Then $du = (1/2\sqrt{x}) dx$ and $dx/\sqrt{x} = 2 \, du$. Substituting we

obtain: $I = 2\int u^2 du = \frac{2}{3} u^3 + C_a = \frac{2}{3}(x^{1/2} + 3)^3 + C_a = \frac{2}{3}(x^{3/2} + 9x + 27x^{1/2} + 27)$

$+ C_a = \frac{2}{3} x^{3/2} + 6x + 18x^{1/2} + 18 + C_a$.

(b) $I = \int \frac{x + 6\sqrt{x} + 9}{\sqrt{x}} dx = \int (\sqrt{x} + 6 + \frac{9}{\sqrt{x}}) dx = \int (x^{1/2} + 6 + 9x^{-1/2}) dx =$

$\frac{2}{3} x^{3/2} + 6x + 18x^{1/2} + C_b$. Comparing answers we see that $C_b = C_a + 18$.

28. Let $u = 3x+2$, $du = 3 \, dx$ so that $dx = (1/3) du$. Then $D_x \int (3x+2)^7 dx =$

$D_x \int \frac{u^7}{3} du = D_x (\frac{u^8}{24} + C) = D_x (\frac{(3x+2)^8}{24} + C) = \frac{8(3x+2)^7 (3)}{24} = (3x+2)^7$.

31. An antiderivative of $D_x \sqrt{x^2+16}$ is $\sqrt{x^2+16}$ itself. Thus $\int_{0}^{3} D_x \sqrt{x^2+16} \, dx =$

$\sqrt{x^2+16}]_{0}^{3} = \sqrt{9+16} - \sqrt{16} = 5 - 4 = 1$.

34. Since $f(x)$ is ≥ 0 and continuous on $[1,2]$, the desired area is $A = \int_{1}^{2} f(x) dx$

$= \int_{1}^{2} (x^2+1)^{-2} x \, dx$. Let $u = x^2+1$, $du = 2x \, dx$, $x \, dx = (1/2) du$. $x = 1 \Rightarrow$

$u = 2$, $x = 2 \Rightarrow u = 5$. Thus $A = \frac{1}{2} \int_{2}^{5} u^{-2} \, du = -\frac{1}{2} u^{-1}]_{2}^{5} = -\frac{1}{2}(\frac{1}{5} - \frac{1}{2}) =$

$-\frac{1}{2}(-\frac{3}{10}) = \frac{3}{20}$.

37. Let $u = x+4$, $du = dx$. $x = 0 \Longrightarrow u = 4$. $x = 5 \Longrightarrow u = 9$. $\int_0^5 \sqrt{x+4}\, dx =$

$\int_4^9 u^{1/2}du = \frac{2}{3}u^{3/2}\Big]_4^9 = \frac{2}{3}(9^{3/2} - 4^{3/2}) = \frac{2}{3}(3^3 - 2^3) = \frac{38}{3}$. Thus we seek

$z \in (0,5)$ (<u>NOT</u> $(4,9)$) such that $\sqrt{z+4}\,(5-0) = \frac{38}{3} \Longrightarrow \sqrt{z+4} = \frac{38}{15} \Longrightarrow z+4 =$

$(\frac{38}{15})^2 \Longrightarrow z = (\frac{38}{15})^2 - 4 \approx 2.418$.

EXERCISES 5.7, page 265

NOTE: T will denote the trapezoidal approximation and S the Simpson Rule approximation. In the tables, m and r denote the coefficients in T and S, respectively.

1. Here $f(x) = 1/x$, $b = 4$, $a = 1$, $n = 6$, $(b-a)/n = .5$. Our work is arranged in the following table.

i	x_i	$f(x_i)$	m	$mf(x_i)$	r	$rf(x_i)$
0	1.0	1.0000	1	1.0000	1	1.0000
1	1.5	0.6667	2	1.3334	4	2.6668
2	2.0	0.5000	2	1.0000	2	1.0000
3	2.5	0.4000	2	0.8000	4	1.6000
4	3.0	0.3333	2	0.6666	2	0.6666
5	3.5	0.2857	2	0.5714	4	1.1428
6	4.0	0.2500	1	0.2500	1	0.2500
				5.6214		8.3262

(a) $(b-a)/2n = 1/4 \Longrightarrow$ T $= 5.6214/4 = 1.40535 \approx 1.41$.

(b) $(b-a)/3n = 1/6 \Longrightarrow$ S $= 8.3262/6 = 1.3877 \approx 1.39$.

4. Tabulating as above, we have $f(x) = \sqrt{1+x^2}$, $(b-a)/n = .25$.

i	x_i	$f(x_i)$	m	$mf(x_i)$	r	$rf(x_i)$
0	2.00	3.0000	1	3.0000	1	3.0000
1	2.25	3.5200	2	7.0400	4	14.0800
2	2.50	4.0774	2	8.1548	2	8.1548
3	2.75	4.6687	2	9.3374	4	18.6748
4	3.00	5.2915	1	5.2915	1	5.2915
				32.8237		49.2011

Since $(b-a)/2n = 1/8$, T $= 32.8237/8 = 4.1030 \approx 4.10$.

Since $(b-a)/3n = 1/12$, S $= 49.2011/12 = 4.1001 \approx 4.10$.

7. For notational convenience, let A $= \int_1^{2.7} \frac{1}{x}\, dx$ and B $= \int_1^{2.8} \frac{1}{x}\, dx$. The trape-

zoidal approximations to A and B are 0.9940 and 1.0304, respectively. (Details omitted.) Since $f'(x) = -x^{-2}$ and $f''(x) = 2x^{-3}$, it follows that $|f''(x)| \le 2$ if

$x \geq 1$. Thus the error in the approximation to A is $\leq \frac{1.7}{12} (.1)^2 (2) = .017/6$ $\leq .003$. Similarly the error in the approximation to B is $\leq \frac{1.8}{12} (.1)^2 (2) =$ $.003$. Thus $0.9910 \leq A \leq 0.9970$; $1.0274 \leq B \leq 1.0334$, and $A < 1 < B$ as desired.

10. From the given data, b = 4, a = 2, n = 10, (b-a)/n = 0.2. We tabulate our work as before.

i	x_i	$f(x_i)$	m	$mf(x_i)$	r	$rf(x_i)$
0	2.0	12.1	1	12.1	1	12.1
1	2.2	11.4	2	22.8	4	45.6
2	2.4	9.7	2	19.4	2	19.4
3	2.6	8.4	2	16.8	4	33.6
4	2.8	6.3	2	12.6	2	12.6
5	3.0	6.2	2	12.4	4	24.8
6	3.2	5.8	2	11.6	2	11.6
7	3.4	5.4	2	10.8	4	21.6
8	3.6	5.1	2	10.2	2	10.2
9	3.8	5.9	2	11.8	4	23.6
10	4.0	5.6	1	5.6	1	5.6
				146.1		220.7

Since (b-a)/2n = .1, T = (.1)(146.1) ≈ 14.61.

Since (b-a)/3n = .2/3, S = .2/3(220.7) ≈ 14.71.

EXERCISES 5.8 (Review), page 266

1. $\sum\limits_{k=1}^{5} (k^2+3) = (1^2+3) + (2^2+3) + (3^2+3) + (4^2+3) + (5^2+3) = 4+7+12+19+28 = 70.$

4. Since [-2,3] is to be partitioned into 5 equal subintervals, each is of length 1 and $x_0 = -2$, $x_1 = -1$, $x_2 = 0$, $x_3 = 1$, $x_4 = 2$, $x_5 = 3$. Since w_i is the midpoint of $[x_{i-1}, x_i]$, $w_1 = -3/2$, $w_2 = -1/2$, $w_3 = 1/2$, $w_4 = 3/2$, $w_5 = 5/2$. Since $\Delta x_i = 1$ for $i = 1, \ldots, 5$, we obtain

$$R_p = f(-3/2) + f(-1/2) + f(1/2) + f(3/2) + f(5/2)$$
$$= (1 - \frac{9}{4}) + (1 - \frac{1}{4}) + (1 - \frac{1}{4}) + (1 - \frac{9}{4}) + (1 - \frac{25}{4})$$
$$= 5 - (9+1+1+9+25)/4 = 5 - 45/4 = -25/4.$$

7. $\int_0^1 \sqrt[3]{8x^7} \, dx = \int_0^1 2x^{7/3} \, dx = 2(\frac{3}{10}) [x^{10/3}]_0^1 = \frac{6}{10} = \frac{3}{5}.$

10. $\int (x^2+4)^2 \, dx = \int (x^4 + 8x^2 + 16) \, dx = \frac{x^5}{5} + \frac{8}{3}x^3 + 16x + C.$

13. Let $t = w^2 + 2w$, $dt = 2(w+1)dw$. When $w = 1$, $t = 3$ and when $w = 2$, $t = 8$.

Thus $\int_1^2 \frac{w+1}{\sqrt{w^2+2w}} dw = \frac{1}{2} \int_3^8 \frac{1}{\sqrt{t}} dt = \frac{1}{2} \int_3^8 t^{-1/2} dt = t^{1/2}\Big]_3^8 = \sqrt{8} - \sqrt{3} \approx 2.828 - $

$1.732 \approx 1.10$.

16. Noting that $(x^2-x-6)/(x+2) = x-3$ if $x \neq -2$, $\int_1^2 \frac{x^2-x-6}{x+2} dx = \int_1^2 (x-3)dx = $

$\frac{x^2}{2} - 3x\Big]_1^2 = -3/2$.

19. Let $u = x^3+1$, $du = 3x^2 dx$. When $x = 0$, $u = 1$, and when $x = 2$, $u = 9$. Then

$\int_0^2 x^2\sqrt{x^3+1}\, dx = \frac{1}{3} \int_1^9 u^{1/2}\, du = \frac{2}{9}u^{3/2}\Big]_1^9 = \frac{2}{9}[9^{3/2} - 1] = \frac{52}{9}$.

22. Let $u = 1 - v^{-1}$, $du = v^{-2}dv$ and $\int \frac{\sqrt[4]{1-v^{-1}}}{v^2} dv = \int \sqrt[4]{u}\, du = \frac{4}{5}u^{5/4} + C = $

$\frac{4}{5}(1 - v^{-1})^{5/4} + C$.

25. The answer includes the constant C because the most general antiderivative of $D_y \sqrt[5]{y^4+2y^2+1}$ is sought.

28. (a) Tabulating our work as before with $f(x) = \sqrt{1+x^4}$ and $(b-a)/n = 10/5 = 2$, we have:

i	x_i	$f(x_i)$	m	$mf(x_i)$
0	0	1.0000	1	1.0000
1	2	4.1231	2	8.2462
2	4	16.0312	2	32.0624
3	6	36.0139	2	72.0278
4	8	64.0078	2	128.0156
5	10	100.0050	1	100.0050
				341.3570

Since $(b-a)/2n = 1$, T = 341.36.

(b) Now, $(b-a)/n = 10/8 = 1.25$.

i	x_i	$f(x_i)$	r	$rf(x_i)$
0	0.00	1.0000	1	1.0000
1	1.25	1.8551	4	7.4204
2	2.50	6.3295	2	12.6590
3	3.75	14.0980	4	56.3920
4	5.00	25.0200	2	50.0400
5	6.25	39.0753	4	156.3012
6	7.50	56.2589	2	112.5178
7	8.75	76.5690	4	306.2760
8	10.00	100.0050	1	100.0050
				802.6114

Since $(b-a)/3n = 10/24$, $S = 8026.114/24 \approx 334.42$.

CHAPTER 6

APPLICATIONS OF THE DEFINITE INTEGRAL

NOTE: In addition to presenting the solutions of #1, 4, 7, etc., I have included
the integral forms of the answers to the remaining odd-numbered problems. In that
wasy, you can determine if the reason for not getting the answer in the back of
the text is due to incorrect set-up (i.e. the wrong integral) or to an arithmetic
or algebraic mistake in evaluating the correct integral.

EXERCISES 6.1, page 278

1. The upper boundary of the region is $y = 1/x^2$; the lower is $y = -x^2$. (See

    sketches in the text for odd-numbered problems.) Thus $A = \int_1^2 [\frac{1}{x^2} - (-x^2)]\,dx$

    $= \int_1^2 (x^{-2} + x^2)\,dx = -x^{-1} + x^3/3\,]_1^2 = 17/6.$

3. $A = \int_{-1}^2 ((4+y) - (-y^2))\,dy$.

4. Because of the shape and boundaries of the
    region, the easiest method is to use y as
    independent variable.

    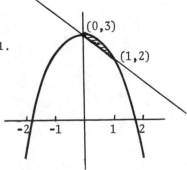

    $A = \int_{-2}^3 [y^2-(y-2)]\,dy$

    $= \frac{y^3}{3} - \frac{y^2}{2} + 2y\,]_{-2}^3$

    $= \frac{21}{2} + \frac{26}{3} = \frac{115}{6}$.

5. $A = \int_{-2}^2 (5 - (x^2+1))\,dx$.

7. For the points of intersection, we solve $x^2 = 4x \iff x^2-4x = x(x-4) = 0 \iff$

    $x = 0,4$. Since $4x \geq x^2$ on $[0,4]$, $A = \int_0^4 (4x-x^2)\,dx = 2x^2 - \frac{x^3}{3}\,]_0^4 = 32 - \frac{64}{3} = \frac{32}{3}.$

9. $A = \int_{-2}^1 ((1 + x^2) - (x-1))\,dx$.

10. The curves intersect when $3-x^2 = -x+3 \iff x = 0,1$.

    $A = \int_0^1 [(3-x^2)-(-x+3)]\,dx = \int_0^1 (-x^2+x)\,dx =$

    $-\frac{x^3}{3} + \frac{x^2}{2}\,]_0^1 = \frac{1}{6}$.

    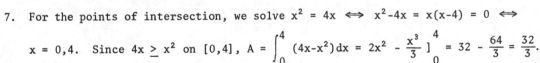

11. $A = \displaystyle\int_{-\sqrt{3}}^{\sqrt{3}} ((2 - y^2) - (y^2 - 4))\,dy$.

13. The lines $y = x$ and $y = 3x$ intersect at $(0,0)$; $y = x$ and $x+y = 4$ intersect at $(2,2)$; $y = 3x$ and $x+y = 4$ intersect at $(1,3)$. The region must be subdivided no matter which variable is chosen as independent. For $0 \le x \le 1$, the upper boundary is $y = 3x$; the lower is $y = x$. For $1 \le x \le 2$, the upper boundary is $x+y = 4$, the lower is $y = x$.

$$A = \int_0^1 (3x-x)\,dx + \int_1^2 [(-x+4)-x]\,dx = \int_0^1 2x\,dx + \int_1^2 (4-2x)\,dx = x^2\Big]_0^1 + 4x-x^2\Big]_1^2$$

$$= 1 + [(8-4)-(4-1)] = 1 + 4 - 3 = 2.$$

15. $A = \displaystyle\int_{-1}^{0} (x^3 - x)\,dx - \int_{0}^{1} (x^3 - x)\,dx$.

16. Writing $y = x(x^2-x-6) = x(x-3)(x+2)$, we see that $y \ge 0$ if $-2 \le x \le 0$ and $y \le 0$ if $0 \le x \le 3$. Thus, using (6.2) on $[0,3]$,

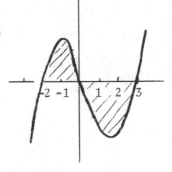

$$A = \int_{-2}^{0} (x^3-x^2-6x)\,dx - \int_{0}^{3} (x^3-x^2-6x)\,dx$$

$$= \left[\frac{x^4}{4} - \frac{x^3}{3} - 3x^2\right]_{-2}^{0} + \left[-\frac{x^4}{4} + \frac{x^3}{3} + 3x^2\right]_{0}^{3}$$

$$= \frac{64}{12} + \frac{63}{4} = \frac{253}{12} .$$

(Note that if you simply integrated y from -2 to 3 without the initial analysis, the "area" would have worked out to be $-125/12$, negative!)

17. $A = -\displaystyle\int_{-2}^{0} (4y - y^3)\,dy + \int_{0}^{2} (4y - y^3)\,dy$.

19. $y = x\sqrt{4-x^2}$ is ≤ 0 if $-2 \le x \le 0$ and is ≥ 0 if $0 \le x \le 2$. Thus

$$A = \int_{-2}^{0} -x\sqrt{4-x^2}\,dx + \int_{0}^{2} x\sqrt{4-x^2}\,dx = \frac{1}{3}(4-x^2)^{3/2}\Big]_{-2}^{0} - \frac{1}{3}(4-x^2)^{3/2}\Big]_{0}^{2} = \frac{2}{3}\cdot 4^{3/2} = \frac{16}{3}.$$

(By symmetry, the area of the right half could have been computed and then doubled to get the answer.)

22. Solving $2x+3y = 6$ for y, we get $y = (-2/3)x + 6$. Let $f(x) = -\frac{2}{3}x + 6$, $0 \le x \le 3$. Then $A = \lim_{\|P\| \to 0} \sum_{i=1}^{n} f(w_i)\Delta x_i$. The region is a triangle of height 2 and base 3 and of area $A = 3$.

23. $A = \displaystyle\int_{0}^{1} (4x + 1)\,dx$.

25. The Δy_i occurring in the sum implies that the in-
     dependent variable is y. Thus $f(w_i) = 4-w_i^2 \Longrightarrow$
     $f(y) = 4-y^2$, $0 \le y \le 1$.
     $$A = \int_0^1 (4-y^2)\,dy = 4y - \frac{y^3}{3}\Big]_0^1 = 4 - \frac{1}{3} = \frac{11}{3}\ .$$

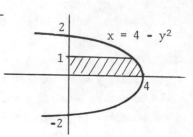

27. $A = \int_2^5 x/(x^2 + 1)^2\,dx\ .$

28. The Δx_i appearing implies that x is the independent variable.

     $f(w_i) = 5w_i/\sqrt{9+w_i^2} \Longrightarrow f(x) = 5x/\sqrt{9+x^2}$, $1 \le x \le 4$, and

     $A = \int_1^4 5x/\sqrt{9+x^2}\ dx.$ Let $u = 9+x^2$, $du = 2x\,dx$, $x\,dx =$

     $du/2.$ $x = 1 \Longrightarrow u = 10.$ $x = 4 \Longrightarrow u = 25.$

     $A = \frac{5}{2} \int_{10}^{25} u^{-1/2}\ du = 5u^{1/2}\Big]_{10}^{25} = 5(\sqrt{25} - \sqrt{10})$

     $= 25 - 5\sqrt{10} \approx 9.2.$

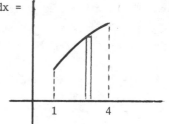

29. $A = \int_1^4 (5 + \sqrt{y})/\sqrt{y}\ dy\ .$

EXERCISES 6.2, page 287

NOTE: When using the disc or washer method to compute the volume of a solid of
revolution, you must remember that the partitioning must be done so that a typical
rectangle is <u>perpendicular</u> to the line around which the revolution is taking place
(the axis of revolution). Thus, if revolution takes place around a vertical line,
a typical rectangle must be horizontal so that a y-interval must be partitioned
and the bounding curves must be graphs of functions of y. Conversely, if we revolve about a horizontal line, the rectangle must be vertical so that an x-interval
must be partitioned and the bounding curves expressed as graphs of functions of x.

1. By (6.5), $V = \pi \int_1^3 (1/x)^2\ dx = \pi \int_1^3 x^{-2}\ dx = \pi[-\frac{1}{x}\Big]_1^3 = \frac{2\pi}{3}\ .$

3. $V = \pi \int_0^2 y\ dy.$

4. Writing the equation as $x = 1/y$, $1 \le y \le 3$, we have by

    (6.6), $V = \pi \displaystyle\int_1^3 (1/y)^2 \, dy = 2\pi/3$.

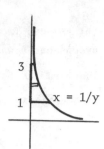

5. $V = \pi \displaystyle\int_0^4 (x^2 - 4x)^2 \, dx$.

7. The curves intersect when $2y = y^2$ or $y = 0,2$. For $0 \le y \le 2$, the right boundary is $x = 2y$; the left is $x = y^2$. Thus a typical rectangle, corresponding to the subinterval $[y_{i-1}, y_i]$ in a partition of $0 \le y \le 2$, sweeps out a "washer" of volume $[\pi(2w_i)^2 - \pi(w_i^2)^2]\Delta y_i$. Summing and passing to the limit as $\|P\| \to 0$, we get

    $$V = \pi \int_0^2 [(2y)^2 - (y^2)^2]\,dy = \pi\left[\frac{4}{3}y^3 - \frac{1}{5}y^5\right]_0^2 = \frac{64\pi}{15}.$$

9. $V = \pi \displaystyle\int_{-\sqrt{2}}^{\sqrt{2}} ((4 - x^2)^2 - x^4)\,dx$.

10. The curves intersect when $x^{1/3} = -x^2 \iff$
    $x = -x^6 \iff x(x^5+1) = 0$ yielding $x = -1,0$.
    Formula (6.7) cannot be used here since the
    functions are both negative for $-1 \le x \le 0$.
    The typical rectangle shown sweeps out a
    washer with outer radius $-w_i^{1/3}$ and inner
    radius w_i^2. (As <u>distances</u>, these radii must
    be ≥ 0.) The washer's volume is
    $\pi[(-w_i^{1/3})^2 - (w_i^2)^2]\Delta x_i$. Summing and
    passing to the limit we obtain

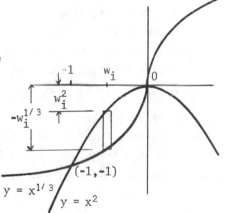

    $$V = \int_{-1}^0 (x^{2/3} - x^4)\,dx = \pi\left[\frac{3}{5}x^{5/3} - \frac{1}{5}x^5\right]_{-1}^0 = \frac{2\pi}{5}.$$

11. $V = \pi \displaystyle\int_{-1}^2 ((y + 2)^2 - y^4)\,dy$.

13. (a) The rectangle shown, when revolved about $y = 4$, sweeps out a disc of base radius $(4-w_i^2)$, thickness Δx_i and volume $\pi(4-w_i^2)^2\Delta x_i$. Thus, as above,

    $$V = \pi \int_{-2}^2 (4-x^2)^2 dx = \pi\left[16x - \frac{8}{3}x^3 + \frac{1}{5}x^5\right]_{-2}^2 = \frac{512\pi}{15}.$$

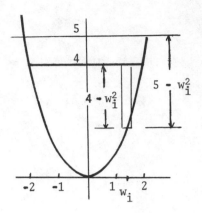

(b) As seen from the sketch, the rectangle now sweeps out a washer of outer radius $5-w_i^2$ and inner radius $5-4 = 1$. Its volume is

$[\pi(5-w_i^2)^2 - \pi]\Delta x_i$. Thus

$$V = \pi\int_{-2}^{2} [(5-x^2)^2-1]\,dx = \pi\int_{-2}^{2} (24-10x^2+x^4)\,dx$$

$$= \pi[24x - \frac{10}{3}x^3 + \frac{1}{5}x^5]\Big|_{-2}^{2} = \frac{832\pi}{15}\ .$$

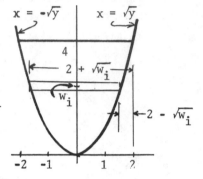

(c) For the revolution about the vertical line $x = 2$, the y-interval, $0 \le y \le 4$, is partitioned. The rectangle shown sweeps out a washer of outer radius $2 + \sqrt{w_i}$, inner radius $2 - \sqrt{w_i}$, and volume $\pi[(2+\sqrt{w_i})^2 - (2-\sqrt{w_i})^2]\Delta y_i$. Thus $V =\pi\int_{0}^{4} [(2+\sqrt{y})^2 -$

$(2-\sqrt{y})^2]\,dy = 8\pi\int_{0}^{4} y^{1/2}\,dy = 8\pi\frac{2}{3}y^{3/2}\Big|_{0}^{4} = \frac{128}{3}\pi.$

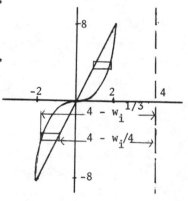

16. The curves $y = x^3$ and $y = 4x$ intersect when $x^3 = 4x$, $x^3 - 4x = x(x^2-4) = 0$, namely, at $x = 0, \pm 2$. The points of intersection are $(-2,-8)$, $(0,0)$, $(2,8)$. Since we are revolving about a vertical line, $x = 4$, we partition the y-interval $[-8,8]$ and express the curves as functions of y:

$x = y^{1/3}$ and $x = y/4$. Because left and right boundaries of the region are different below and above the x-axis we must consider the lower and upper portions of the solid separately. When $[-8,0]$ is partitioned, a typical rectangle sweeps

out a washer with outer radius $4-w_i^{1/3}$ and inner radius $4-w_i/4$. (Since $w_i < 0$, both of these are greater than 4 as the sketch shows.) The volume of this washer is $\pi[(4-w_i^{1/3})^2 - (4-w_i/4)^2]\Delta y_i$. Summing and passing to the limit we find that the lower portion of the solid has volume equal to

$\pi\int_{-8}^{0} [(4-y^{1/3})^2 - (4-y/4)^2]\,dy$. When $[0,8]$ is partitioned, a typical rectangle sweeps out a washer with inner and outer radii interchanged. Thus the volume of the upper portion of the solid is $\pi\int_{0}^{8} [(4-y/4)^2 - (4-y^{1/3})^2]\,dy$, and the total volume is the sum of these two integrals.

19. To use the washer method we partition the y
     interval [-1,1] since revolution is around
     the vertical line x = 5. The left half of
     the circle has equation $x = -\sqrt{1-y^2}$, and the
     right half is $x = \sqrt{1-y^2}$. A typical rectangle,
     as shown, sweeps out a washer of outer radius
     $5 - (-\sqrt{1-w_i^2}) = 5 + \sqrt{1-w_i^2}$ and inner radius
     $5 - \sqrt{1-w_i^2}$ Its volume is $\pi[(5 + \sqrt{1-w_i^2})^2 -$

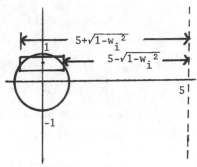

     $(5 - \sqrt{1-w_i^2})^2]\Delta y_i = 20\pi\sqrt{1-w_i^2}\Delta y_i$. Summing and passing to the limit, we ob-

     tain $V = 20\pi\displaystyle\int_{-1}^{1} \sqrt{1-y^2}\,dy$. (Note that the solid is a doughnut, or torus.)

22. To form the sphere, we revolve the semicir-
     cular region under $y = \sqrt{r^2-x^2}$, $-r \le x \le r$,

     about the x-axis. Thus $V = \pi\displaystyle\int_{-r}^{r} y^2\,dx =$

     $\pi\displaystyle\int_{-r}^{r} (r^2-x^2)\,dx = \pi[r^2x - \frac{x^3}{3}]\Big|_{-r}^{r} = \frac{4}{3}\pi r^3$.

25. The Δx_i implies that the interval being partitioned is $0 \le x \le 1$. The term,

     $\pi(w_i^4 - w_i^6)\Delta x_i = \pi[(w_i^2)^2 - (w_i^3)^2]\Delta x_i$, is the volume of a washer of outer

     radius w_i^2 and inner radius w_i^3. The sum and limit then yield the given

     answer. Also: $V = \pi\displaystyle\int_{0}^{1} (x^4-x^6)\,dx = \pi[\frac{x^5}{5} - \frac{x^7}{7}]\Big|_{0}^{1} = \frac{2\pi}{35}$.

EXERCISES 6.3, page 293

NOTE: When using the shell method to compute the volume of a solid of revolution,
the partitioning must be done so that a typical rectangle's altitude is <u>parallel</u>
to the axis of revolution. Thus, revolution about a vertical line means that an
x-interval must be partitioned, the altitude of the rectangle is vertical, and
the bounding curves must be expressed as functions of x. Similarly, revolution
about a horizontal line means a y-interval must be partitioned and the bounding
curves expressed as functions of y.

     It should also be noted that, although many of the solutions below are ob-
tained using analysis of the shell obtained by revolving a typical rectangle, it
is possible to derive integral formulas for volumes. For example, suppose
$f(x) \ge g(x)$ on [a,b], where f and g are continuous, and suppose that the region
between the graphs of $f(x)$ and $g(x)$ from a to b is revolved about the y-axis.
We partition the interval [a,b], and let w_i be the midpoint of $[x_{i-1},x_i]$. When

the rectangle shown is revolved about the y-axis, a
shell is swept out with altitude $f(w_i) - g(w_i)$,
average radius w_i, thickness Δy_i, and volume
$2\pi w_i(f(w_i) - g(w_i))$. The volume of the solid is
approximately the sum of these shell volumes, and,
thus, summing and passing to the limit as $\|P\| \to 0$,

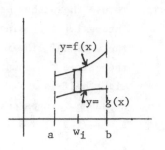

we obtain $V = 2\pi \displaystyle\int_a^b x(f(x) - g(x))\,dx$.

A similar formula can be derived if $f(y) \geq g(y)$ on a y-interval $[c,d]$, if
the region between the graphs is revolved about the x-axis. We obtain:

$$V = 2\pi \int_c^d y(f(y) = g(y))\,dy.$$

1. By (6.10), $V = 2\pi \displaystyle\int_0^4 x\sqrt{x}\,dx = 2\pi \int_0^4 x^{3/2}dx = 2\pi(\tfrac{2}{5})x^{5/2}\Big]_0^4 = \dfrac{4\pi}{5}(2^5) = \dfrac{128\pi}{5}$.

3. $V = 2\pi \displaystyle\int_0^2 x(\sqrt{8x} - x^2)\,dx$.

4. $x^2 - 5x = x(x-5) \leq 0$ on $[0,5]$ since $x \geq 0$,
    $x-5 \leq 0$ there. So, with rotation about the
    y-axis (vertical), we partition this x interval.
    The shell swept out by the rectangle shown has
    altitude $0 - (w_i^2 - 5w_i) = 5w_i - w_i^2$, average
    radius w_i and thickness Δy_i. Its volume is,
    from (6.9), $2\pi w_i(5w_i - w_i^2)$. Summing, and
    passing to the limit as $\|P\| \to 0$ we obtain

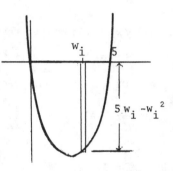

$$V = 2\pi \int_0^5 x(5x-x^2)\,dx = 2\pi \int_0^5 (5x^2-x^3)\,dx = 2\pi[\tfrac{5}{3}x^3 - \tfrac{x^4}{4}]_0^5 = 2\pi[\tfrac{625}{3} - \tfrac{625}{4}] =$$

$\dfrac{1250\pi}{12} = \dfrac{625\pi}{6}$.

5. $V = 2\pi \displaystyle\int_4^7 x[(\tfrac{x}{2} - \tfrac{3}{2}) - (2x - 12)]\,dx$.

7. To go around the x-axis and use the shell method, we partition the y-interval,
    $0 \leq y \leq 4$. Solving $x^2 = 4y$ for x, we find that the left boundary is $x = -2\sqrt{y}$,
    and the right boundary is $x = 2\sqrt{y}$. The typical rectangle is horizontal with
    altitude = (right boundary) - (left boundary) = $2\sqrt{w_i} - (-2\sqrt{w_i}) = 4\sqrt{w_i}$. The
    shell has average radius w_i, thickness Δy_i, and volume $2\pi w_i(4\sqrt{w_i})\Delta y_i =$

$8\pi w_i^{3/2}\Delta y_i$. Summing and passing to the limit, we obtain $V = 8\pi \int_0^4 y^{3/2}dy =$

$\dfrac{16\pi}{5}\, y^{5/2}]_0^4 = \dfrac{16\pi}{5}(32)$.

9. $V = 2\pi \int_0^6 y(y/2)\,dy$.

10. $2y = x$ intersects $x = 1$ at $y = 1/2$. The right
boundary is $x = 2y$; the left is $x = 1$. Thus,
the altitude is $2w_i - 1$, radius w_i, thickness
Δy_i. The shell volume is $2\pi w_i(2w_i-1)\Delta y_i$. Thus,
as above, $V = 2\pi \int_{1/2}^4 y(2y-1)\,dy = 2\pi[\frac{2}{3}y^3 - \frac{y^2}{2}]_{1/2}^4$

$= 2\pi(\dfrac{104}{3} + \dfrac{1}{24}) = \dfrac{833\pi}{12}$.

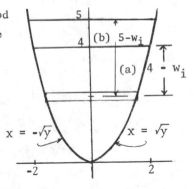

11.   (a) $V = 2\pi \int_0^2 (3-x)(x^2+1)\,dx$.   (b) $V = 2\pi \int_0^2 (x+1)(x^2+1)\,dx$.

13.   (a) To revolve about $y = 4$ and use the shell method
we partition $0 \le y \le 4$. The average radius of the
shell swept out by the rectangle shown is $(4-w_i)$;
the height is (right boundary - left boundary) $=$
$\sqrt{w_i} - (-\sqrt{w_i}) = 2\sqrt{w_i}$, and the volume is
$2\pi(4-w_i) \cdot 2\sqrt{w_i}\Delta y_i$. Summing and passing to the
limit as before, $V = 4\pi \int_0^4 (4-y)y^{1/2}dy = 512\pi/15$.

(b) The analysis is similar to (a). The average radius here is $5-w_i$ and

$V = 4\pi \int_0^4 (5-y)y^{1/2}dy = 832\pi/15$.

(c) Partition the x-interval $[-2,2]$. The shell swept out by a rectangle now
has average radius $(2-w_i)$ and height $(4-w_i^2)$ with volume $2\pi(2-w_i)(4-w_i^2)\Delta x_i$.

Thus $V = 2\pi \int_{-2}^2 (2-x)(4-x^2)\,dx = 128\pi/3$.

16.   Here, we partition x-intervals. On $[-2,0]$, $x^3 \ge 4x$ and the altitude of the
rectangle and shell is $w_i^3 - 4w_i$. The average shell radius is $4-w_i$, thickness
Δx_i, and its volume is $2\pi(4-w_i)(w_i^3 - 4w_i)\Delta x_i$. Summing and passing to the
limit we obtain for the volume of the lower portion of the solid,

$2\pi \int_{-2}^{0} (4-x)(x^3-4x)\,dx$. When $[0,2]$ is partitioned,

$4x \geq x^3$, and the altitude is $4w_i - w_i^3$. The
radius and thickness are the same, and

$2\pi \int_{0}^{2} (4-x)(4x-x^3)\,dx$ is the volume of the upper

portion. The total volume is the sum of these
two integrals.

19. We partition the x-interval $[-1,1]$ and express the upper and lower portions
    of the circle as functions of x, $y = \pm\sqrt{1-x^2}$. The typical rectangle is ver-
    tical with altitude $\sqrt{1-w_i^2} - (-\sqrt{1-w_i^2}) = 2\sqrt{1-w_i^2}$. The average radius is
    $5-w_i$, and the volume of the shell is $2\pi(5-w_i)(2\sqrt{1-w_i^2})\Delta x_i$. Summing and

    passing to the limit, $V = 2\pi \int_{-1}^{1} (5-x)(2\sqrt{1-x^2})\,dx$.

22. To form the sphere, revolve the right half of the circular region $x^2+y^2 \leq r$
    about the y-axis. The upper and lower boundaries are $y = \pm\sqrt{r^2-x^2}$. Thus

    $$V = 2\pi \int_{0}^{r} x[\sqrt{r^2-x^2} - (-\sqrt{r^2-x^2})]\,dx = 4\pi \int_{0}^{r} x(r^2-x^2)^{1/2}\,dx = 4\pi(-\tfrac{1}{3})(r^2-x^2)^{3/2}\Big]_{0}^{r}$$

    $$= \tfrac{4}{3}\pi r^3 .$$

25. The Δx_i again tells us that $0 \leq x \leq 1$ is being partitioned. The term
    $2\pi(w_i^2 - w_i^3)\Delta x_i = 2\pi w_i(w_i-w_i^2)\Delta x_i$ is the volume of a shell of average
    radius w_i, height $(w_i-w_i^2)$, and thickness Δx_i. The sum and limit then yield

    the given answer. $V = 2\pi \int_{0}^{1} (x^2-x^3)\,dx = \pi/6$. (Note that if we wrote the term

    as $\pi[(\sqrt{2}w_i)^2 - (\sqrt{2}w_i^{3/2})^2]\Delta x_i$ the answer could also be the volume of the

    solid obtained by rotating the region between $y = \sqrt{2}x$ and $y = \sqrt{2}x^{3/2}$ about
    the x-axis.)

EXERCISES 6.4, page 297

1. If (x,y) is a point on the circle for which $y \geq 0$ then, because of the symmetry of the circle, the side of the cross-sectional square has length $2y$,
   and the area of the square is $A = 4y^2$. Since $x^2 + y^2 = a^2$, $A = 4(a^2 - x^2)$

   and $V = \int_{-a}^{a} 4(a^2-x^2)\,dx = 4[a^2x - \frac{x^3}{3}]_{-a}^{a} = \frac{16a^3}{3}$.

3. $V = \int_{-2}^{2} [(4-x^2)/2]^2 dx.$

4. If (x,y) is on the curve $y = x^2$, the cross-sectional
    square has side length $(4-y) = 4-x^2$ and area
    $A = (4-x^2)^2$. Thus

    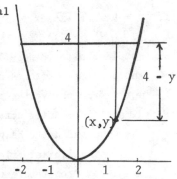

    $$V = \int_{-2}^{2} (4-x^2)^2 \, dx = \int_{-2}^{2} (16 - 8x^2 + x^4) \, dx$$

    $$= [16x - \frac{8}{3}x^3 + \frac{1}{5}x^5]_{-2}^{2} = 2 \left[32 - \frac{64}{3} + \frac{32}{5} \right] = \frac{512}{15} .$$

5. $V = \int_{0}^{h} (2a^2x^2/h^2) \, dx.$

7. If (x,y) is on $y^2 = 4x$, the base (diameter) of the
    cross-sectional semi-circle has length $4-x =$
    $4 - y^2/4$. The radius is $(4-y^2/4)/2$ and the area of
    the semi-circle is $A(y) = \frac{\pi}{8}(4 - \frac{y^2}{4})^2$ and

    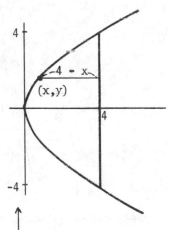

    $$V = \frac{\pi}{8} \int_{-4}^{4} (4 - \frac{y^2}{4})^2 \, dy = \frac{\pi}{8} \int_{-4}^{4} (16 - 2y^2 + \frac{y^4}{16}) \, dy$$

    $$= \frac{\pi}{8}[16y - \frac{2}{3}y^3 + \frac{y^5}{80}]_{-4}^{4} = \frac{128\pi}{15} .$$

9. $V = 2\int_{0}^{a} y\sqrt{a^2 - y^2} \, dy.$

10. Position the x-axis with the origin at
    the point of intersection of the cylin-
    drical axes and perpendicular to the
    plane they lie in. The sketch depicts
    only 1/8 of the entire solid, the nearest
    1/4 of the top half.

    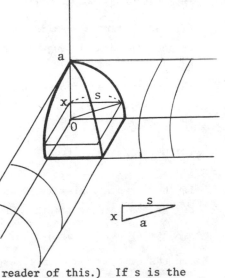

    Restricting our attention to this por-
    tion, we see that for a number $x \in [0,a]$,
    the plane through x perpendicular to the
    x-axis cuts the solid in a square. (The
    facts that the edges of the cut are on
    the cylindrical surface parallel to the
    axis of the cylinder, and that the cylin-
    ders are perpendicular should convince the reader of this.) If s is the
    length of a side, then $s = \sqrt{a^2-x^2}$ (see insert in sketch). The area of the
    square is $A(x) = s^2 = a^2-x^2$ and the volume is

$\int_0^a (a^2 - x^2) dx = \frac{2}{3}a^3$. The entire solid therefore has volume $8(\frac{2}{3})a^3 = \frac{16a^3}{3}$.

11. $V = b\int_{-a}^{a} \sqrt{a^2 - x^2} \ dx$.

13. We place a coordinate line along the 4 cm. side with the origin at the vertex
shown. (See sketches at end of solution.) Consider a point on this line x
units from the vertex; a plane through this point, perpendicular to the coordinate line, intersects the solid in a triangular cross-section. Let b be
the base and h the height of this triangle. We can express b and h in terms
of x using similar triangles. From the middle sketch below, b/x = 2/4, or
b = x/2. From the last sketch, h/x = 3/4, or h = (3/4)x. The area of this
cross-sectional triangle is A(x) = (1/2)bh = 1/2 · x/2 · 3/4x = $\frac{3}{16}$ x^2 . Thus

$$V = \int_0^4 (3/16)x^2 dx = \frac{1}{16}x^3 \Big]_0^4 = \frac{64}{16} = 4.$$

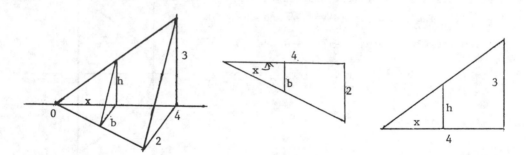

EXERCISES 6.5, page 304

1. If, as in Example 2, x is the number of units that the spring is stretched
beyond its natural length, then f(x) = kx and f(1.5) = 8 $\Longrightarrow$ k(3/2) = 8 $\Longrightarrow$
k = 16/3.

(a) Here x ranges from 0 to 4 (14" is 4" beyond the natural length of 10").

Thus $W = \int_0^4 (16/3)x \ dx = 128/3$.

(b) Here, x ranges from 1 to 3 and $W = \int_1^3 (16/3)x \ dx = \frac{64}{3}$.

4. In the formula f(x) = kx, x is the amount that the spring has been stretched
beyond its natural length. Thus, if x_0 is the natural length, then we obtain
amounts of stretch of $6-x_0$, $7-x_0$, and $8-x_0$ when the spring is stretched to
lengths of 6, 7 and 8 cm., respectively. The data given in the problem,
together with (6.13) and (6.14) yield: $\int_{6-x_0}^{7-x_0} kx \ dx = 60$ and $\int_{7-x_0}^{8-x_0} kx \ dx = 120$.

We seek k and x_o. From the first relation we get: $(k/2)[(7-x_o)^2-(6-x_o)^2]$

$= 60 \implies k(13-2x_o) = 120 \implies 13k - 2kx_o = 120$. From the second relation,

we get: $(k/2)[(8-x_o)^2-(7-x_o)^2] = 120 \implies k(15-2x_o) = 240 \implies 15k - 2kx_o =$

240. Subtracting from this equation the equation from the first relation, we

get $2k = 120$ or $k = 60$. Using this in either equation, we get $x_o = 11/2 =$

5.5 cm.

5. $W = 62.5 \displaystyle\int_0^3 8(3-y)\,dy.$

7. By (6.12), the work done in lifting the 3000 lb.
    elevator 10' is 30,000 ft-lb. To this we must
    add the work of lifting the cable. We place
    the y-axis with the origin at the initial posi-
    tion of the bottom of the cable and $y = 10$ at
    the final position. Partition [0,10] and let
    $w_i \in [y_{i-1},y_i]$. When the cable bottom is at w_i,
    there are $12-w_i$ ft. of cable still suspended with

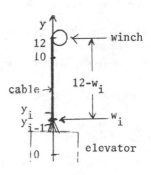

a mass of $15(12-w_i)$. The work done in lifting

the cable from y_{i-1} to y_i is thus $\approx 15(12-w_i)\Delta y_i$. The total work in lifting

the cable is approximately the sum of all such terms. In the limit as

$\|P\| \to 0$ we get $\displaystyle\int_0^{10} 15(12-y)\,dy = 15(120-50) = 1050$ ft. lb. for the work in

lifting the cable. Thus the total work in lifting the cable and elevator is
31,050 ft. lb.

9. (a) $W = 62.5\pi(9/4) \displaystyle\int_0^6 (6-y)\,dy.$ (b) $W = 62.5\pi(9/4) \displaystyle\int_0^6 (10-y)\,dy.$

10. Placing the axes as shown we partition the y-interval,
    $0 \le y \le 3$, since the tank is only half filled. The ith
    slice has volume $V_i = \pi(3/2)^2\Delta y_i$ and mass $62.5\,V_i =$

    $(562.5\pi/4)\Delta y_i.$

    (a) The work required to lift this slice to the top is
    (the distance lifted) · (mass) $\approx (6-w_i)$ · (mass). The

    total work is approximately the sum of these, and,
    passing to the limit as $\|P\| \to 0$ we obtain

    $W = \displaystyle\int_0^3 (6-y)(562.5\pi/4)\,dy = (562.5\pi/4)(27/2) \approx 5,964$ ft. lbs.

(b) The only difference from (a) is that the slice must be lifted $(10-w_i)$

ft. Thus $W = \int_0^3 (10-y)(562.5\pi/4)\,dy = (562.5\pi/4)(51/2) \approx 11{,}266$ ft.lbs.

11. $W = \dfrac{16(62.5)}{\sqrt{3}} \displaystyle\int_0^{\sqrt{3}} y(\sqrt{3} - y)\,dy.$

13. Here $f(x) = k/d^2$ where k is a constant, and d is the distance between the electrons.

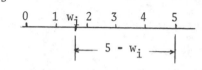

(a) Partition $0 \le x \le 3$. At $w_i \,\varepsilon\, [x_{i-1}, x_i]$,

$d = 5-w_i$, $f(w_i) = k/(5-w_i)^2$. The work done

in moving the electron from x_{i-1} to x_i is approximately (force) · (distance)

$= \dfrac{k\Delta x_i}{(5-w_i)^2}$. The total work done in moving it from $x = 0$ to $x = 3$ is ap-

proximately the sum of all such terms, and passing to the limit as $\|P\| \to 0$,

we obtain $W = \displaystyle\int_0^3 \dfrac{k}{(5-x)^2}\,dx = \dfrac{k}{5-x}\Big]_0^3 = k(\tfrac{1}{2} - \tfrac{1}{5}) = \dfrac{3k}{10}$.

(b) Begin as in (a). Now, the net force, f, on the electron at w_i is a sum $f_1 + f_2$ of the force f_1 exerted by the electron at

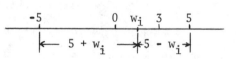

$(-5,0)$ and the force f_2 exerted by the elec-

tron at (5.0). Note that f_1 and f_2 must be of opposite signs since f_2 op-

poses the motion (as in (a)), but f_1 assists it in repelling the moving elec-

tron. Thus $f(w_i) = \dfrac{-k}{(5+w_i)^2} + \dfrac{k}{(5-w_i)^2}$ and, as above,

$W = \displaystyle\int_0^3 [-\dfrac{k}{(5+x)^2} + \dfrac{k}{(5-x)^2}]\,dx = k[\dfrac{1}{5+x} + \dfrac{1}{5-x}]_0^3 = \dfrac{9k}{40}$.

15. $W = 115 \displaystyle\int_{32}^{40} v^{-1.2}\,dv.$

16. Since $pv^{1.14} = c$ for all values of p and v, we can use the initial values to

find $c = p_0 v_0^{1.14}$. Thus using $p = cv^{-1.14}$ and Example 4 we obtain

$W = \displaystyle\int_{v_0}^{2v_0} cv^{-1.14}\,dv = \dfrac{c}{-0.14} v^{-0.14}\Big]_{v_0}^{2v_0} = cv_0^{-0.14}(1 - 2^{-0.14})/0.14$, which

when we substitute the value of c, becomes $p_0 v_0(1-2^{-0.14})/0.14 \approx 0.66 p_0 v_0$.

19. Here $a = 0$, $b = 5$, and n, the number of subintervals (one less than the number

of points), is 10. Using (5.47), $W = \int_0^5 f(x)\,dx \approx \frac{5-0}{2(10)} [7.4 + 2(8.1) + 2(8.4)$

$+ 2(7.8) + 2(6.3) + 2(7.1) + 2(5.9) + 2(6.8) + 2(7.0) + 2(8.0) + 9.2] =$
$0.25(147.4) = 36.8$.

EXERCISES 6.6, page 309

NOTE: In these solutions, ρ is the density of water, 62.5 lbs/ft^3.

1. (a) Introducing a coordinate system with the origin at the lower left corner,
    we can use (6.16) directly with k = 1, f(y) = 1, g(y) = 0 so that F =

    $\rho \int_0^1 (1-y)(1-0)\,dy = \rho/2$.

    (b) Similar to (a) except that f(y) = 3 and $F = \rho \int_0^1 (1-y)(3-0)\,dy = 3\rho/2$.

2. HINT: Let the diagonal be the line y = x. For the upper half, f(y) = y,
    g(y) = 0, k = 1. For the lower half, f(y) = 1, g(y) = y, k = 1.

3. (a) $F = \rho \int_0^1 (1-y)2\sqrt{3}\ y\ dy$.   (b) $F = \rho \int_0^{1/2} (1/2-y)2\sqrt{3}\ y\ dy$.

4. The end of the trough and the coordinate axes
    are shown. (The length is immaterial.) The
    upper vertices are at $x = \pm\sqrt{4-h^2}$. Thus the
    right boundary is $y = (h/\sqrt{4-h^2})x$ or $x =$
    $(\sqrt{4-h^2}/h)y = f(y)$. Similarly the left boun-
    dary is $x = (-\sqrt{4-h^2}/h)y$.

    (a) The trough full means the surface is at

    $y = h$. Thus $F = \rho \int_0^h (h-y)[(\sqrt{4-h^2}/h)(y-(-y))]\ dy = 2\rho(\sqrt{4-h^2}/h)\int_0^h (h-y)y\ dy$.

    The integral is $h^3/6$ and $F = \rho h^2\sqrt{4-h^2}/3$.
    (b) If half full, the surface is at y = h/2, and $F = 2\rho(\sqrt{4-h^2}/h)\int_0^{h/2} (h/2-y)y\,dy$.
    The integral is $h^3/48$ and $F = \rho h^2\sqrt{4-h^2}/24$.

5. $F = 60\int_{-2}^0 (0-y)\ 2\sqrt{4-y^2}\ dy$.

7. Place the x-axis on a sloping edge of the bottom with x = 0 at the 3' end and
    $x = \sqrt{40^2 + 6^2} = \sqrt{1636}$ at the 9' end. Partition as usual and consider the
    rectangle shown. Above the point w_i, the depth d is 3 + L where, by similar
    triangles, $\frac{L}{6} = \frac{w_i}{\sqrt{1636}}$ so that $d = 3 + \frac{6w_i}{\sqrt{1636}}$. The force acting on this
    rectangle = (pressure) $\cdot$ (area) $\approx (\rho d) \cdot (20\Delta x_i)$. Summing over all

rectangles and passing to the limit, the

total force is $F = 20\rho \int_0^{\sqrt{1636}} (3 + \frac{6x}{\sqrt{1636}}) dx$

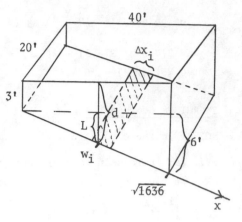

$= 120\sqrt{1636}\, \rho \approx 4853.7\rho$. (Another way would
be to place the axis on a horizontal edge.
With a bit of work, the area at the bottom
can be shown to be $\sqrt{1636}\Delta x_i/2$ and the depth

at w_i is $3 + \frac{6w_i}{40}$.)

9. $F = \rho \int_0^4 (10-y)[(4-y/2)-(y/2-4)]\,dy$.   (The y-axis bisects the trapezoid for

   this form of the integral.)

10. Place the coordinate axes with the origin at the
    center of the circle, the water level at $y = 6$.
    The right and left boundaries of the circular
    plate are $f(y) = \sqrt{4-y^2}$ and $g(y) = -\sqrt{4-y^2}$. Thus

    $F = 2\rho \int_{-2}^2 (6-y)\sqrt{4-y^2}\,dy$. To evaluate $\int_{-2}^2 \sqrt{4-y^2}\,dy$,

    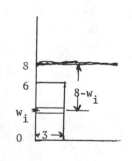

    note that this is the area of half a circle of
    radius 2, which is 2π. The other integral is

    $\int_{-2}^2 y\sqrt{4-y^2}\,dy = -\frac{1}{3}(4-y^2)^{3/2}\Big]_{-2}^2 = 0$. Thus $F = 24\pi\rho$.

11. $F = \rho \int_0^4 (4-y)\, 2\sqrt{y}\,dy$.

13. Place the y axis with $y = 0$ at the bottom of the plate.
    Then the top is at $y = 6$, and the surface is at $y = 8$.
    Partition $[0,6]$ and let $w_i \varepsilon [y_{i-1}, y_i]$. By (6.17) the

    force on the typical rectangle shown is $50(8-w_i)(3\Delta y_i)$

    $= 150(8-w_i)\Delta y_i$. Summing and passing to the limit as

    $\|P\| \to 0$ we get $F = 150 \int_0^6 (8-y)\,dy = 150(48-18) = 4500$ lbs.

14. HINT. Let the diagonal extend from the lower left to the upper right corner.
    For the left part of the plate, by similar triangles, the area of the typical
    rectangle is $(3/6)w_i\Delta y_i$.

EXERCISES 6.7, page 316

1. $8x^2 = 27y^3 \implies y = (2/3)x^{2/3} \implies y' = (4/9)x^{-1/3} = f'(x)$. Then $\sqrt{1 + f'(x)^2}$

$= \sqrt{1 + (16/81)x^{-2/3}} = \sqrt{x^{-2/3}(x^{2/3} + 16/81)} = x^{-1/3}\sqrt{x^{2/3} + 16/81}$, and $L_1^8 =$

$\int_1^8 x^{-1/3}\sqrt{x^{2/3} + 16/81}\ dx$. Let $u = x^{2/3} + 16/81$ so that $du = (2/3)x^{-1/3}\ dx$

or $x^{-1/3}\ dx = (3/2)du$. $x = 1 \implies u = 1 + 16/81 = A$. $x = 8 \implies u = 8^{2/3} +$

$16/81 = 4 + 16/81 = B$. Thus $L_1^8 = (3/2)\int_A^B u^{1/2}\ du = u^{3/2}\ ]_A^B = B^{3/2} - A^{3/2} =$

$(4 + 16/81)^{3/2} - (1 + 16/81)^{3/2} \approx 7.29$.

3. $L_1^4 = \int_1^4 \sqrt{1 + (9/4)x}\ dx$.

4. Since $y = 6x^{2/3} + 1$, the calculations are similar to those in #1. The only

thing to watch out for is that $\sqrt{x^{2/3}} = -x^{1/3}$ in this problem since $-8 \le x \le -1$

< 0 here. Then $L_{-8}^{-1} = \int_{-8}^{-1} -x^{-1/3}\sqrt{x^{2/3} + 16}\ dx = 20^{3/2} - 17^{3/2}$ with the sub-

stitution $u = x^{2/3} + 16$. (See #10 also.)

5. $L_{2/3}^{8/3} = \int_{2/3}^{8/3} \sqrt{1 + (243/32)y}\ dy$.

7. $y' = \dfrac{x^2}{4} - \dfrac{1}{x^2} = f'(x)$. Then $1 + f'(x)^2 = 1 + (\dfrac{x^4}{16} - \dfrac{1}{2} + \dfrac{1}{x^4}) = \dfrac{x^4}{16} + \dfrac{1}{2} + \dfrac{1}{x^4} =$

$(\dfrac{x^2}{4} + \dfrac{1}{x^2})^2$. Thus, $L_1^2 = \int_1^2 (\dfrac{x^2}{4} + x^{-2})\ dx = \dfrac{x^3}{12} - \dfrac{1}{x}\ ]_1^2 = \dfrac{1}{6} - (-\dfrac{11}{12}) = \dfrac{13}{12}$.

9. $L_1^2 = \int_1^2 (\dfrac{y^4}{6} + \dfrac{3}{2}y^{-4})\ dy$ (by (6.19)).

10. $x = g(y) = \dfrac{y^4}{16} + \dfrac{1}{2y^2} \implies g'(y) = \dfrac{y^3}{4} - \dfrac{1}{y^3}$. Then $1 + g'(y)^2 = 1 + (\dfrac{y^6}{16} - \dfrac{1}{2} + \dfrac{1}{y^6})$

$= \dfrac{y^6}{16} + \dfrac{1}{2} + \dfrac{1}{y^6} = (\dfrac{y^3}{4} + \dfrac{1}{y^3})^2$. Here, $-2 \le y \le -1 < 0$ so that $y^3 < 0$ and the ex-

pression in the parentheses is negative. Recalling that $\sqrt{a^2} = |a| = -a$ if

$a < 0$, it follows that $\sqrt{1 + g'(y)^2} = -(\dfrac{y^3}{4} + \dfrac{1}{y^3})$. Thus, $L_{-2}^{-1} =$

$-\int_{-2}^{-1} (\dfrac{y^3}{4} + y^{-3})\ dy = -[\dfrac{y^4}{16} - \dfrac{1}{2y^2}]_{-2}^{-1} = -(-\dfrac{7}{16} - \dfrac{7}{8}) = \dfrac{21}{16}$.

13. Let the line $y = x$ and the given curve intersect when $x = a$. Solving, we

find that $2a^{2/3} = 1$ so that $a = (1/2)^{3/2}$. By symmetry then, the desired

length is $8\ L_a^1$. Solving $x^{2/3} + y^{2/3} = 1$ for y, we get $y = (1-x^{2/3})^{3/2}$ and

$y' = -(1-x^{2/3})^{1/2}/x^{1/3}$ and $\sqrt{1+y'^2} = \sqrt{1+(1-x^{2/3})/x^{2/3}}$

which simplifies to $x^{-1/3}$. Thus $8\,L_a^1 =$

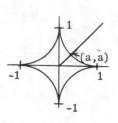

$$8\int_a^1 x^{-1/3}\,dx = 12(1 - a^{2/3}) = 12(1 - 1/2) = 6.$$

NOTE. If we had tried to compute L_0^1 directly, we would

have obtained $L_0^1 = \int_0^1 1/\sqrt[3]{x}\,dx$ which does not exist be-

cause the integrand becomes infinite as $x \to 0^+$. (See pp. 233-234 of the
text.) This type of integral is called an "improper" integral and will be
studied in Chapter 11.

16. $f(x) = x^{3/2} \Longrightarrow f'(x) = (3/2)x^{1/2} \Longrightarrow \sqrt{1 + f'(x)^2} = \sqrt{1 + 9x/4}$. Thus $s(x) =$

$\int_1^x \sqrt{1 + 9t/4}\,dt = \frac{8}{27}\,[(1 + 9t/4)^{3/2}\,]_1^x = \frac{8}{27}\,[(1 + \frac{9}{4}x)^{3/2} - (1 + \frac{9}{4})^{3/2}] =$

$\frac{1}{27}\,[(4 + 9x)^{3/2} - 13^{3/2}]$. $\Delta s = s(1.1) - s(1) = (\frac{1}{27})[13.9^{3/2} - 13^{3/2}] \approx$

$(51.82 - 46.86)/27 \approx 0.184$.

ds $= \sqrt{1 + f'(x)^2}\,\Delta x$ and with $x = 1$, $\Delta x = 0.1$, ds $= \sqrt{1 + 9/4}\,(.1) = \sqrt{13}(.1)/2$

≈ 0.180.

19. $f'(x) = 3x^2 \Longrightarrow L_0^2 = \int_0^2 \sqrt{1 + 9x^4}\,dx$. Let $g(x) = \sqrt{1 + 9x^4}$ so that $\int_0^2 g(x)\,dx$

must be estimated. By (5.49), with $a = 0$, $b = 2$ and $n = 4$ we have:

$\int_0^2 g(x)\,dx \approx \frac{2-0}{3(4)}\,[g(0) + 4g(\frac{1}{2}) + 2g(1) + 4g(\frac{3}{2}) + g(2)] =$

$(1/6)[1.000 + 4(1.250) + 2(3.162) + 4(6.824) + 12.042] = 8.610$.

EXERCISES 6.8, page 322

1. The total depreciation, $f(t)$, is $\int_0^t g(x)\,dx = \int_0^t (1-x^2/9)\,dx = t - t^3/27$ hun-
    dreds of dollars.
    (a) At 6 months, $t = 1/2$ and $f(1/2) = 1/2 - 1/8(27) = 107/216 \approx \$49.54$.
    (b) $f(1) = 26/27 \approx \$96.30$.   (c) $f(3/2) = 11/8 \approx \$137.50$.
    (d) $f(2) = 46/27 \approx \$170.37$.

4. Over $[0,5]$, the amount of capital formation is $\int_0^5 2t(3t+1)\,dt = \int_0^5 (6t^2+2t)\,dt$

    $= 275$ thousand dollars. Over $[5,10]$, we obtain

    $\int_5^{10} (6t^2+2t)\,dt = 2100-275 = 1825$ thousands of dollars.

7. (a) 1-item time is $f(1) \approx 18.16$.   (b) 4-item time $\approx \int_0^4 (20(x+1)^{-.4}+3)\,dx =$

$\frac{20}{.6}$ $(x+1)^{.6}]_0^4$ $+12 = \frac{20}{.6}(5^{.6}-1)+12 \approx 66.22$.   (c) 8-item time $\approx \frac{20}{.6}(9^{.6}-1)+32 \approx$

115.24.   (d) 16-item time $\approx \frac{20}{.6}(17^{.6}-1)+48 \approx 197.12$ min.

EXERCISES 6.9 (Review), page 323

1. The curves intersect at $(\pm 2,4)$. Thus for x integration the interval is
    $-2 \le x \le 2$, the upper boundary is $y = -x^2$, the lower is $y = x^2 - 8$, and

    $A = \int_{-2}^{2} [(-x^2)-(x^2-8)]\,dx = \int_{-2}^{2} (8-2x^2)\,dx = 64/3$. For the y integration the

    interval is $-8 \le y \le 0$. For $-8 \le y \le -4$, the right and left boundaries are
    $x = \pm(y+8)^{1/2}$, and for $-4 \le y \le 0$, the boundaries are $x = \pm(-y)^{1/2}$. Thus

    $A = \int_{-8}^{-4} 2(y+8)^{1/2}\,dy + \int_{-4}^{0} 2(-y)^{1/2}\,dy = \frac{32}{3} + \frac{32}{3} = \frac{64}{3}$.

3. $A = \int_a^b [(1-y)-y^2]\,dy$ where $a = (-1 - \sqrt{5})/2$, $b = (-1 + \sqrt{5})/2$.

4. Find point A by substituting $y = -x^3$ into the
    line's equation to obtain $-3x^3 + 7x - 10 = 0$
    which has $x = -2$ as the only real solution.
    For point B, put $y = \sqrt{x}$, or $y^2 = x$, into the
    linear equation to obtain $3y + 7y^2 - 10 = 0$
    with positive solution $y = 1$. Thus the x-
    interval is $-2 \le x \le 1$ with the obvious change
    in the lower boundary at $x = 0$. Thus

    $A = \int_{-2}^{0} [(-\frac{7}{3}x + \frac{10}{3}) - (-x^3)]\,dx +$

    $\int_{0}^{1} [(-\frac{7}{3}x + \frac{10}{3}) - x^{1/2}]\,dx = \frac{1}{3} \int_{-2}^{1} (-7x+10)\,dx + \int_{-2}^{0} x^3\,dx - \int_{0}^{1} x^{1/2}\,dx =$

    $\frac{27}{2} - 4 - \frac{2}{3} = \frac{53}{6}$.

5. $V = \pi \int_0^2 (4x + 1)\,dx$.

7. Using the shell method, $V = 2\pi \int_0^1 x(2-(x^3+1)\,dx = 2\pi \int_0^1 (x-x^4)\,dx = 3\pi/5$.

9. (a) $V = \pi \int_{-2}^{1} [(-4x+8)^2 - (4x^2)^2]\,dx$.

    (b) $V = 2\pi \int_{-2}^{1} (1-x)[(-4x+8) - 4x^2]\,dx$.

(c) $V = \pi \int_{-2}^{1} [(16-4x^2)^2 - (4x+8)^2] dx.$

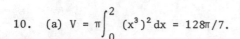

10. (a) $V = \pi \int_{0}^{2} (x^3)^2 dx = 128\pi/7.$

(b) $V = 2\pi \int_{0}^{2} x(x^3) dx = 64\pi/5.$

(c) The indicated rectangle sweeps out a cy-lindrical shell of average radius $(2-w_i)$ and volume $2\pi(2-w_i)w_i^3 \Delta x_i$. Thus

$$V = 2\pi \int_{0}^{2} (2-x)x^3 dx = 2\pi[x^4/2 - x^5/5] \Big|_{0}^{2} = 16\pi/5.$$

(d) This is similar to (c) except that the average radius is $(3-w_i)$. Thus

$$V = 2\pi \int_{0}^{2} (3-x)x^3 dx = 56\pi/5.$$

(e) The rectangle now sweeps out a washer of outer radius 8, inner radius $(8-w_i^3)$, and volume $\pi[8^2-(8-w_i^3)^2]\Delta x_i$. Thus $V =$

$$\pi \int_{0}^{2} [64-(8-x^3)^2] dx = \pi \int_{0}^{2} (16x^3-x^6) dx = \pi[4x^4-x^7/7] \Big|_{0}^{2} = 320\pi/7.$$

(f) Now the washer has outer radius $1+w_i^3$, inner radius 1, and volume $\pi[(1+w_i^3)^2 - 1^2]\Delta x_i$. Thus $V = \pi \int_{0}^{2} [(1+x^3)^2 - 1] dx = \pi \int_{0}^{2} (x^6+2x^3) dx = 184\pi/7.$

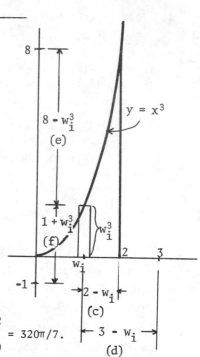

11. $L_{-2}^{5} = \int_{-2}^{5} \sqrt{(x+3)^{2/3} + 9} \ (x+3)^{-1/3} \ dx.$

13. With the axes as shown, partition $0 \le y \le 4$. At height w_i, the volume of the ith slice is $\pi \cdot 6^2 \cdot \Delta y_i$. The mass is $36\pi\rho\Delta y_i$ and the work done in lifting it to the top of the pool $(y=5)$ is $(5-w_i)\cdot$(mass). Summing and taking the limit as $\|P\| \to 0$ we obtain

$$W = 36\pi\rho \int_{0}^{4} (5-y) dy = 432\pi\rho \text{ ft.lbs.}$$

15. $F = \rho \displaystyle\int_{0}^{2\sqrt{2}} (6-y)2(2\sqrt{2} - y)\,dy + \rho \int_{-2\sqrt{2}}^{0} (6-y)2(2\sqrt{2} + y)\,dy.$ (The y-axis bisects

the plate for this form of the answer.)

16. $y = 3x^2 - 4x + 2 \implies y' = 6x - 4 \implies ds = \sqrt{1 + (6x-4)^2}\ \Delta x =$
$\sqrt{17 + 36x^2 - 48x}\ \Delta x.$ With $x = 1$ and $\Delta x = .1$, $\Delta s \approx ds = \sqrt{17 + 36 - 48}$ (.1)
$= \sqrt{5}(.1) \approx .224.$

CHAPTER 7

TOPICS IN ANALYTIC GEOMETRY

EXERCISES 7.2, page 333

1. $x^2 = -12y$ is in standard form for a parabola with
    vertex $(0,0)$ and focus $(0,p)$ where $4p = -12$ or
    $p = -3$. Thus, focus $(0,-3)$ and directrix $y = 3$.

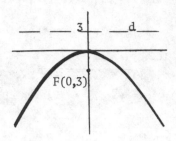

2. For $y^2 = x/2$ we have $V(0,0)$ and $F(p,0)$ where
    $4p = 1/2$ or $p = 1/8$. Thus $F(1/8,0)$ and directrix
    is $x = -1/8$.

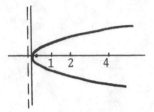

4. Done as #1 with $4p = -3$, $F(0,-3/4)$, directrix $y = 3/4$.

7. $y^2 - 12 = 12x \iff y^2 = 12(x+1)$. Thus $V(-1,0)$, $4p = 12$
    or $p = 3 > 0$, and it opens to the right. Thus the
    focus is $F(-1+3,0) = (2,0)$ and the directrix is
    $x = -1 - 3 = -4$.

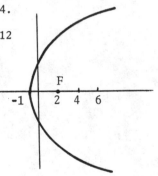

10. Completing the square: $y = 8x^2 + 16x + 10 = 8(x^2 + 2x + 1)$
     $+ 2$ yields $(x+1)^2 = (y-2)/8$. Thus $V(-1,2)$, $4p = 1/8$ or
     $p = 1/32 > 0$ and the parabola opens upward with
     $F(-1,2 + 1/32) = (-1,65/32)$ and directrix $y = 2 - 1/32 =$
     $63/32$.

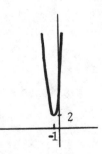

13. Divide by 4 and rewrite to obtain $x^2 + 10x = -\frac{1}{4}y - \frac{106}{4}$. Completing the
     square: $x^2 + 10x + 25 = -\frac{1}{4}y - \frac{106}{4} + 25$, or $(x+5)^2 = -\frac{1}{4}y - \frac{6}{4} = -\frac{1}{4}(y+6)$.
     Thus $V(-5,-6)$, $4p = -\frac{1}{4}$, or $p = -\frac{1}{16}$ so the parabola opens downward with
     $F(-5,-6 - \frac{1}{16}) = F(-5,-\frac{97}{16})$ and directrix $y = -6 + \frac{1}{16} = -\frac{95}{16}$.

13. (Graph)

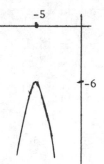

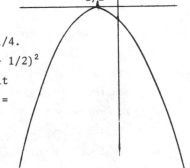

16. Divide by 4 and rewrite to obtain $x^2 + x = -y - 1/4$.
     Completing the square: $x^2 + x + 1/4 = -y$ or $(x + 1/2)^2$
     $= -y$. Thus $V(-1/2,0)$, $4p = -1$, or $p = -1/4$, so it
     opens downward with $F(-1/2,-1/4)$ and directrix $y =$
     $-1/4$.

19. With directrix $x = -2$, it opens to the left or right and therefore has equa-
     tion $(y-k)^2 = 4p(x-h)$, vertex (h,k), focus $(h+p,k)$, directrix: $x = h-p$.
     Using the given data we obtain the system of equations $h+p = 2$, $k = 0$, $h-p =$
     -2. Solving, we get $h = 0$, $p = 2$ and so the equation is $y^2 = 8x$, opening to
     the right.

22. With directrix $y = 1$, it opens up or down and has equation $(x-h)^2 = 4p(y-k)$,
     vertex (h,k), focus $(h,k+p)$, directrix $y = k-p$. Using the data given we ob-
     tain the system: $h = -3$, $k+p = -2$, $k-p = 1$. Solving, we get $k = -1/2$, $p =$
     $-3/2 < 0$. Thus it opens downward and has equation $(x+3)^2 = 4(-3/2)(y + 1/2)$
     $= -6(y + 1/2)$.

25. Place the coordinate axes with the origin at the vertex of
     the cross-sectional parabola and with this curve opening to
     the right so that its equation is $y^2 = 4px$ $(p > 0)$. Given
     that when $x = 1$, $2y = 3$ or $y = 3/2$ we have: $(3/2)^2 = 4p(1)$
     or $p = 9/16$. Thus the focus is at $(9/16,0)$, i.e. 9/16 feet
     from the vertex of the parabolic reflector.

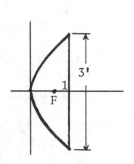

31. $x^2 = 4y \implies p = 1$ and the focus is $(0,1)$; the line, ℓ, is $y = 1$.

     (a) $A = \int_{-2}^{2} (1 - \frac{x^2}{4})\,dx = [x - \frac{x^3}{12}]_{-2}^{2} = \frac{8}{3}$.

     (b) To obtain the solid, we rotate the right half of the region, $0 \le x \le 2$, about the y-axis, and, so, by the shell method

$$V = 2\pi \int_0^2 x(1 - \frac{x^2}{4})\,dx = 2\pi\,[\frac{x^2}{2} - \frac{x^4}{16}]\,\Big]_0^2 = 2\pi.$$

(c) By the "washer" method, $V = \pi \int_{-2}^2 [1^2 - (\frac{x^2}{4})^2]\,dx$

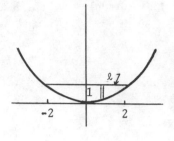

$$= \pi \int_{-2}^2 (1 - \frac{x^4}{16})\,dx = \pi\,[x - \frac{x^5}{80}]\,\Big]_{-2}^2 = \frac{16\pi}{5}.$$

34. There are several proofs--here's one sugges-
    tion. Find the slope of ℓ by implicit
    differentiation and, after writing down
    the equation of ℓ, find the x-intercept,
    Q. The angle between ℓ and QF is equal to
    β (parallel lines cut by a transversal).
    Show that $d(Q,F) = d(P,F)$ so that triangle
    QPF is isoceles and, thus $\beta = \alpha$.

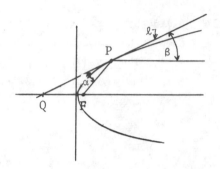

EXERCISES 7.3, page 339

1. This is a standard form (7.9) with $a^2 = 9$, $b^2 = 4$ and
   $c^2 = a^2 - b^2 = 5$. Thus vertices $(\pm 3,0)$, foci $(\pm\sqrt{5},0)$

4. Rewriting the equation as $x^2 + y^2/9 = 1$, this is in
   standard form (7.11) with $a^2 = 9$, $b^2 = 1$ and $c^2 = 8$.
   Thus vertices $(0,\pm 3)$, foci $(0,\pm 2\sqrt{2})$.

7. Rewriting as $x^2/(1/4) + y^2/(1/25) = 1$, this is (7.9)
   with $a^2 = 1/4$, $b^2 = 1/25$, $c^2 = 1/4 - 1/25 = 21/100$.
   Thus vertices $(\pm 1/2,0)$, foci $(\pm\sqrt{21}/10,0)$.

10. Vertices $(0,\pm 7) \implies$ the equation is of the form (7.11), $x^2/b^2 + y^2/a^2 = 1$
    with $a = 7$. The foci $(0,\pm 2) \implies c = 2$. Thus $b^2 = a^2 - c^2 = 49 - 4 = 45$ and
    the equation is $x^2/45 + y^2/49 = 1$.

13. Similar to #10, but now $a = 6$. Thus the equation is $x^2/b^2 + y^2/36 = 1$.
    Since $(3,2)$ is on the ellipse, $x = 3$, $y = 2$ must satisfy this equation.
    Thus $9/b^2 + 4/36 = 1$ or $9/b^2 = 8/9$, which yields $b^2 = 81/8$, which produces
    the answer given.

16. Subtract the second equation from the first to get $3y^2 = 24$ or $y = \pm\sqrt{8} =$
    $\pm\ 2\sqrt{2}$. Substitute $y^2 = 8$ in either equation to get $x^2 = 4$ or $x = \pm 2$.

Thus we have 4 points: $(2, \pm 2\sqrt{2})$, $(-2, \pm 2\sqrt{2})$.

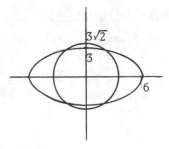

19. $5x^2 + 4y^2 = 56 \Longrightarrow 10x + 8yy' = 0 \Longrightarrow y' = -5x/4y$. At $(-2,3)$, $y' = 5/6$ which yields the equation $y-3 = (5/6)(x+2)$.

22. If (x_1,y_1) is an unknown point of tangency, then, by Exercise 20, the tangent line is $xx_1/a^2 + yy_1/b^2 = 1$. Since $(0,d)$ is on this tangent line by assumption, its coordinate must satisfy its equation. Thus $dy_1/b^2 = 1$ or $y_1 = b^2/d$. Using the equation of the ellipse, $x_1^{\;2} = a^2 - a^2b^2/d^2$ and $x_1 = \pm a\sqrt{1 - b^2/d^2}$. Thus the points of tangency: $(\pm a\sqrt{1-b^2/d^2}, \; b^2/d)$.

25. Major axis of length $16 = 2a \Longrightarrow a = 8$; minor axis of length $9 = 2b \Longrightarrow b = 9/2$. Let us position the coordinate axes with the major axis of the ellipse along the x-axis. Then the equation of the ellipse is $x^2/64 + 4y^2/81 = 1$.
(a) If (x,y) is on the ellipse, with $y > 0$, then the side of the square cross section is of length $2y$, and its area is $4y^2 = 81(1 - x^2/64)$. Thus $V =$

$$81\int_{-8}^{8} (1 - x^2/64)\,dx = 81(32/3) = 864.$$

(b) With (x,y) as above, the base of the cross-sectional equilateral triangle is $2y$ and the area is $\sqrt{3y^2} = \frac{\sqrt{3}}{4}(2y)^2$. Thus the integrand for V is $\sqrt{3}/4$ times the integrand in part (a). Thus $V = (\sqrt{3}/4)864 = 216\sqrt{3}$.

28. Placing the origin at the center, the given data produce the equation $x^2/9 + y^2/4 = 1$ for the edge of the tank. The water extends from $y = -2$ to $y = 0$ since it's half full. The right and left boundaries are $x = \pm(3/2)\sqrt{4 - y^2}$ and, since the surface is at $y = 0$, we have $F = \rho\int_{-2}^{0} (0-y)\cdot$

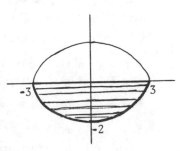

$3\sqrt{4-y^2}\,dy = \rho[(4- y^2)^{3/2}]\Big|_{-2}^{0} = \rho\cdot 8 = 500$ lb.

EXERCISES 7.4, page 345

1. The equation is in standard form (7.15) with
    $a^2 = 9$, $b^2 = 4$. Thus $a = 3$, $b = 2$, $c^2 =$
    $9+4 = 13$, $c = \sqrt{13}$. These data yield the
    answer given.

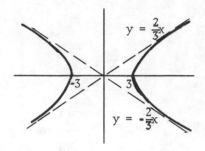

4. This is similar to #1 with $a = 7$, $b = 4$, $c^2 = 49+16 = 65$. Thus the vertices
    $V(\underline{+}7,0)$, foci $F(\underline{+}\sqrt{65},0)$, and asymptotes $y = \underline{+}(4/7)x$. The graph is like the
    above but with a different scale.

5. Rewriting in standard form: $y^2/16 - x^2/4 = 1$.
    Thus $a = 4$, $b = 2$, $c = \sqrt{20}$ yielding the answer
    given.

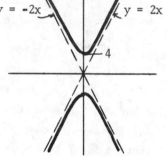

7. Same as #1 and #4 with $a = b = 1$.

10. Rewriting in standard form we have $\dfrac{y^2}{1/4} - \dfrac{x^2}{1/4} = 1$. Thus $a^2 = b^2 = 1/4 \Longrightarrow$
    $a = b = 1/2$ and $c = \sqrt{1/4 + 1/4} = \sqrt{1/2} = 1/\sqrt{2}$. $\Longrightarrow$ $V(0,\underline{+}1/2)$, $F(0,\underline{+}1/\sqrt{2})$,
    and asymptotes $y = \underline{+}x$.

13. F and V on the y-axis implies that the hyperbola opens up and down as in
    Figure 7.20. The equation is $y^2/a^2 - x^2/b^2 = 1$. From $V(0,\underline{+}1)$ we have $a = 1$,
    and from $F(0,\underline{+}4)$ we have $c = 4$. Thus $b^2 = c^2 - a^2 = 16-1 = 15$ yielding
    $y^2 - x^2/15 = 1$ as one form of the equation.

16. F and V on the y-axis $\Longrightarrow$ the hyperbola opens up and down and has equation
    $y^2/a^2 - x^2/b^2 = 1$. The given data give $a = 2$, $c = 3$. Thus $b^2 = c^2 - a^2$
    $= 9-4 = 5$ and the equation is $y^2/4 - x^2/5 = 1$.

19. V on the x-axis $\Longrightarrow$ it opens to the left and right with equation $x^2/a^2 -$
    $y^2/b^2 = 1$ and asymptotes $y = \underline{+}(b/a)x$. From V, $a = 3$. Since the slope of the
    asymptotes is $\underline{+}2 = \underline{+}(b/a) = \underline{+}(b/3)$, we get $b = 6$. Thus the equation:
    $x^2/9 - y^2/36 = 1$.

22. Add the equations to get $x^2-3x = 4 \Rightarrow x = 4,-1$.
    The solution $x = 4$ yields $y^2 = 12$, $y = \pm 2\sqrt{3}$. The
    solution $x = -1$ yields no points of intersection
    since $x \geq 0$ on the parabola.

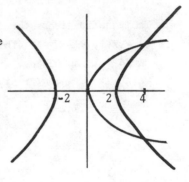

25. $2x^2-5y^2 = 3 \Rightarrow 4x-10yy' = 0 \Rightarrow y' = 2x/5y$. At
    $(-2,1)$, $y' = -4/5$. Thus: $(y-1) = (-4/5)(x+2)$.

28. If (x_1,y_1) is a point of tangency, then, by Exercise 20, the equation of the
    tangent line is $xx_1/a^2 - yy_1/b^2 = 1$. Since $(0,d)$ is on such a tangent line
    by hypothesis, its coordinates must satisfy this equation. Thus $-dy_1/b^2 = 1$
    or $y_1 = -b^2/d$. Substituting this into the equation of the hyperbola, we get
    $x_1^2 = a^2(1 + y_1^2/b^2) = a^2(1 + b^4/d^2b^2) = a^2(1+b^2/d^2)$. Thus the points of
    tangency are $(\pm a\sqrt{1 + b^2/d^2},-b^2/d)$.

31. Using the disc method, $V = \pi\int_a^c y^2 \, dx = b^2\pi\int_a^c (\frac{x^2}{a^2} - 1) dx = \pi b^2[\frac{c^3-a^3}{3a^2} - (c-a)]$

    $= \frac{\pi b^2(c-a)}{3a^2} [(c^2+ac+a^2) - 3a^2]$, which reduces to $\frac{\pi b^2}{3a^2} [c(b^2-2a^2) + 2a^3]$ using
    $c^2 = a^2 + b^2$.

EXERCISES 7.5, page 351

NOTE: In this section F', V', M', and C' will denote the focus, vertex (or vertices), ends of the minor axis, and center of a conic in the translated x'y'-
coordinate system. The same symbols with no primes denote the same points but in
the original xy system. The coordinates in each system are, of course, related
by formulas (7.19) and (7.20).

1. The equation is of the form $y'^2 = 8x'$ where $x' = x+1$,
   $y' = y-5$ (i.e. $h = -1$, $k = 5$). Thus the graph is
   a parabola with V'(0,0) and F'(2,0), ($4p = 8$, so
   $p = 2$). Thus: V(-1,5), F(1,5).

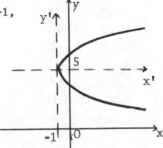

4. The equation is of the form $4x'^2 + 8y'^2 = 32$ where
   $x' = x+1$, $y' = y-5$. This reduces to $x'^2/8 +$
   $y'^2/4 = 1$, an ellipse with $C'(0,0)$, $V'(\pm\sqrt{8},0)$,
   $M'(0,\pm2)$. Thus $C(-1,5)$, $V(-1\pm\sqrt{8},5)$, $M(-1,5\pm2)$.

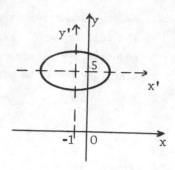

7. This is of the form $4x'^2 + y'^2 = 1$ where $x' = x+5$, $y' =$
   $y-3$ (i.e. $h = -5$, $k = 3$). This reduces to $x'^2/(1/4) +$
   $y'^2 = 1$, an ellipse with $C'(0,0)$, $V'(0,\pm1)$, $M'(\pm1/2,0)$.
   Thus $C(-5,3)$, $V(-5,3\pm1)$, $M(-5\pm1/2,3)$.

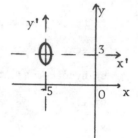

10. This is of the form $x'^2 - y'^2 = 1$ where $x' = x-5$,
    $y' = y+1$ (i.e. $h = 5$, $k = -1$). Thus the graph
    is a hyperbola with $C'(0,0)$, $V'(\pm1,0)$, and asymp-
    totes $y' = \pm x'$. Thus $C(5,-1)$, $V(5\pm1,-1)$, and
    asymptotes $y+1 = \pm(x-5)$.

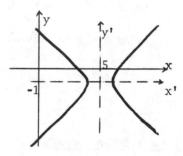

13. Rewriting as $25(x^2 - 10x\quad) + (y^2 - 4y\quad)$
    $= -541$, complete the square in each parentheses to
    obtain $25(x-5)^2 + 4(y-2)^2 = -541 + 625 + 16$ which
    reduces to $x'^2/4 + y'^2/25 = 1$ where $x' = x-5$, $y' =$
    $y-2$, an ellipse with $C'(0,0)$, $V'(0,\pm5)$, $M'(\pm2,0)$.
    Thus $C(5,2)$, $V(5,2\pm5)$, $M(5\pm2,2)$.

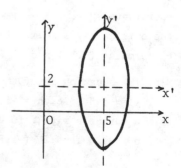

16. Rewriting as $4(y^2 - 4y\quad) = x-13$ and completing
    the square, we get $4(y-2)^2 = x+3$, or $y'^2 = x'/4$
    where $y' = y-2$, $x' = x+3$, a parabola opening to
    the right with $V'(0,0)$, $F'(1/16,0)$ or $V(-3,2)$,
    $F(-47/16,2)$.

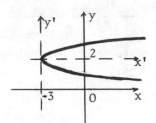

19. Completing the square, we get $9(y^2-4y\quad) -$
    $(x^2-12x\quad) = 36$, $9(y-2)^2 - (x-6)^2 = 36 + 36 - 36$
    or $9y'^2 - x'^2 = 36$ where $x' = x-6$, $y' = y-2$. This
    reduces to $y'^2/4 - x'^2/36 = 1$, a hyperbola with
    $C'(0,0)$, $V'(0,\pm2)$, and asymptotes $y' = \pm x'/3$.
    Thus $C(6,2)$, $V(6,2\pm2)$, and asymptotes $y-2 =$
    $\pm(x-6)/3$.

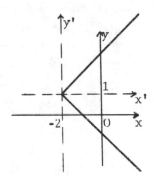

22. This is the same as $x+2 = |y-1|$ or $x' = |y'|$ where
    $x' = x+2$, $y' = y-1$. The graph consists of the two
    half-lines $x' = y'$ $(y' \geq 0)$ and $x' = -y'$ $(y' \leq 0)$.
    Thus $x+2 = y-1$ (if $y \geq 1$), and $x+2 = -y+1$ (if
    $y \leq 1$).

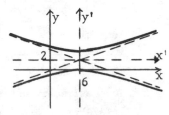

25. This yields $2(y+5) = (x+2)^4$ or, $2y' = x'^4$
    where $y' = y+5$, $x' = x+2$.

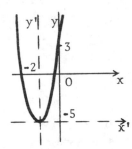

28. Translate the origin to (h,k). Then $x' = x-h$, $y' =$
    $y-k$. Then $V'(0,\pm a)$ and the asymptotes are $y' =$
    $\pm(a/b)x'$. Thus the equation is $y'^2/a^2 - x'^2/b^2$
    $= 1$ or $(y-k)^2/a^2 - (x-h)^2/b^2 = 1$.

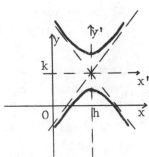

EXERCISES 7.6, page 356

NOTE: The primed notation will be used again in this section where it will denote points in the rotated system. The formulas relating x,y to x',y' are (7.26) and (7.27).

1. cot 2ϕ = (32-53)/(-72) = 21/72. Thus $0 < 2\phi < \pi/2$. Since cot 2ϕ = cos 2ϕ/sin 2ϕ, we have sin 2ϕ = (72/21)cos 2ϕ and, using $\sin^2 2\phi + \cos^2 2\phi = 1$, it follows that $\cos^2 2\phi(72^2/21^2 + 1) = 1$ or $\cos^2 2\phi = 21^2/(72^2 + 21^2) = 21^2/75^2$ $\implies$ cos 2ϕ = 21/75.

Thus sin $\phi = \sqrt{\dfrac{1 - (21/75)}{2}} = \sqrt{\dfrac{54}{150}} = \dfrac{3}{5}$ and

cos $\phi = \sqrt{\dfrac{1 + (21/75)}{2}} = \sqrt{\dfrac{96}{150}} = \dfrac{4}{5}$.

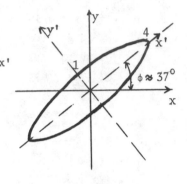

The desired transformation formulas are x = (4/5)x' - (3/5)y', y = (3/5)x' + (4/5)y'. Substituting into the original equation: (32/35)(4x'-3y')^2 - (72/25)(4x'-3y')(3x'+4y') + (53/25)(3x'+4y')^2 = 80, which, after considerable work reduces to $125x'^2 + 2000y'^2 = 2000$ or $x'^2 + 16y'^2 = 16$, an ellipse with C'(0,0), V'($\pm$4,0), M'(0,$\pm$1). Thus C(0,0), V($\pm$16/5,$\pm$12/5), M($\pm$3/5,$\mp$4/5).

4. cot 2ϕ = (1-1)/1 = 0. Thus $2\phi = \pi/2$, $\phi = \pi/4$ and the rotation formulas are x = (1/$\sqrt{2}$)(x'-y'), y = (1/$\sqrt{2}$)(x'+y'). Substituting: (1/2)(x'-y')^2 - (1/2)(x'-y')(x'+y') + (1/2)(x'+y')^2 = 3 which reduces to $x'^2 + 3y'^2 = 6$, an ellipse with C'(0,0), V'($\pm\sqrt{6}$,0), M'(0,$\pm\sqrt{2}$). Thus C(0,0), V($\pm\sqrt{3}$,$\pm\sqrt{3}$), M($\pm$1, $\mp$1).

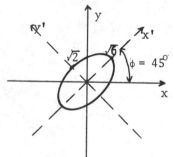

7. cot 2ϕ = (16-9)/(-24) = -7/24. Thus $\pi/2 < 2\phi$ $< \pi$, and cos $2\phi < 0$, sin $2\phi > 0$. Proceeding as in Exercise 1 above, we find that sin 2ϕ = 24/25, cos 2ϕ = -7/25, sin $\phi = \sqrt{(1+7/25)/2}$ = 4/5, cos $\phi = \sqrt{(1-7/25)/2}$ = 3/5. Thus the rotation formulas are: x = (1/5)(3x'-4y'), y = (1/5)(4x'+3y'). Substitution into the original equation eventually yields $y'^2 - 4x' + 4 = 0$ or $y'^2 = 4(x'-1)$, a parabola with V'(1,0), F'(2,0). Thus V(3/5,4/5) and F(6/5,8/5).

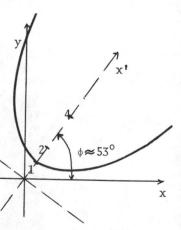

10. $\cot 2\phi = (18-82)/(-48) = 4/3 \implies \sin 2\phi = (3/4)\cos 2\phi$.

$\sin^2 2\phi + \cos^2 2\phi = 1 \implies [(9/16) + 1]\cos^2 2\phi = 1 \implies$

$\cos^2 2\phi = 16/25 \implies \cos 2\phi = 4/5$. Thus $\sin \phi =$

$\sqrt{(1 - 4/5)/2} = \sqrt{1/10} = 1/\sqrt{10}$, and $\cos \phi = \sqrt{(1 + 4/5)/2}$

$= \sqrt{9/10} = 3/\sqrt{10}$. The transformation formulas are

$x = (1/\sqrt{10})(3x'-y')$, $y = (1/\sqrt{10})(x'+3y')$. (This cor-

responds to a counterclockwise rotation of $\phi \approx 18°26'$.)

Substituting these relations into the original equation

yields: $(18/10)(3x'-y')^2 - (48/10)(3x'-y')(x'+3y') +$

$(82/10)(x'+3y')^2 + 6(3x'-y') + 2(x'+3y') - 80 = 0$. When expanded this reduces

to: $10x'^2 + 90y'^2 + 20x' - 80 = 0$, or, $x'^2 + 2x' + 9y'^2 = 8$. Completing the

square: $(x'^2 + 2x' + 1) + 9y'^2 = 9$, or $(x'+1)^2/9 + y'^2 = 1$. Thus, the graph in

the rotated coordinate system is an ellipse with $C'(-1,0)$, $V'(2,0)$ and $V'(-4,0)$,

$M'(-1,1)$ and $M'(-1,-1)$, $F'(-1+\sqrt{8},0)$.

13. $\cot 2\phi = (40-25)/(-36) = -5/12 \implies \cos 2\phi = -5/13$ and $\sin 2\phi = 12/13$.

$\sin \phi = \sqrt{(1 + 5/13)/2} = \sqrt{18/26} = 3/\sqrt{13}$ and $\cos \phi = \sqrt{(1 - 5/13)/2} = 2/\sqrt{13}$.

(Thus $\phi \approx 56°19'$.) The transformation equations are $x = (1/\sqrt{13})(2x'-3y')$,

$y = (1/\sqrt{13})(3x'+2y')$. Substituting into the original equation and combining

terms yields $x'^2 - 4x' + 4y'^2 = 0$. Completing the square: $(x'^2 - 4x' + 4)$

$+ 4y'^2 = 4$, or $(x'-2)^2/4 + y'^2 = 1$ which yields the answer given in the text.

16. The work is set up in the following table:

Ex	A	B	C	B^2-4AC	Graph
1	32	-72	53	-1600	ellipse
2	7	-48	-7	2500	hyperbola
3	11	$10\sqrt{3}$	1	256	hyperbola
4	1	-1	1	-3	ellipse
5	5	-8	5	-36	ellipse
6	11	$-10\sqrt{3}$	1	256	hyperbola
7	16	-24	9	0	parabola
8	1	$2\sqrt{3}$	3	0	parabola
9	5	$6\sqrt{3}$	-1	128	hyperbola
10	18	-48	82	-3600	ellipse
11	1	4	4	0	parabola
12	15	20	$-4\sqrt{5}$	$80(5+3\sqrt{5})$	hyperbola
13	40	-36	25	-2704	ellipse
14	64	-240	225	0	parabola

EXERCISES 7.7 (Review), page 357

1. This is the standard form of a parabola opening to the right with $V(0,0)$. Since $4p = 64$ or $p = 16$, the focus is $F(16,0)$.

4. $9y^2 = 144 + 16x^2 \implies y^2/16 - x^2/9 = 1$, a hyperbola opening up and down with vertices $V(0,\pm 4)$ (since $a^2 = 16$), foci $F(0,\pm 5)$ since $c^2 = 16 + 9 = 25$.

7. $25y = 100 - x^2 \implies x^2 = -25y + 100 = -25(y-4)$, by (7.5), a parabola opening downward since $4p < 0$. We have $V(0,4)$, $4p = -25$, or $p = -25/4$ so that $F(0,4-25/4) = F(0,-9/4)$ and directrix $y = 4 + 25/4 = 41/4$.

10. $x = 2y^2 + 8y + 3 \implies y^2 + 4y = x/2 - 3/2 \implies y^2 + 4y + 4 = x/2 + 5/2$, or $(y+2)^2 = (1/2)(x+5)$, a parabola opening to the right (since $4p = 1/2 > 0$) with $V(-5,-2)$, $p = 1/8$ so that $F(-5+1/8,-2) = F(-39/8,-2)$ and directrix $x = -5 - 1/8 = -41/8$.

13. Since the vertex is equidistant from the focus and directrix, the given data implies $V(0,0)$. Also, $p = -10$ and it opens downward. Thus $x^2 = -40y$.

16. The hyperbola opens to the left and right since F and V are on the x-axis. $V(\pm 5,0) \implies a = 5$; $F(\pm 10,0) \implies c = 10$, and $b^2 = c^2 - a^2 = 75$. Thus, equation: $x^2/25 - y^2/75 = 1$.

19. $4x^2 + 9y^2 + 24x - 36y + 36 = 0 \iff 4(x^2+6x+9) + 9(y^2-4y+4) = -36 + 36 + 36 \iff 4x'^2 + 9y'^2 = 36$ where $x' = x+3$, $y' = y-2$, an ellipse with $C'(0,0)$, $V'(\pm 3,0)$, $M'(0,\pm 2)$. Thus $C(-3,2)$, $V(-3\pm 3,2)$, $M(-3,2\pm 2)$.

22. $4x^2 + y^2 - 24x + 4y + 36 = 0 \iff 4(x^2-6x+9) + (y^2+4y+4) = -36 + 36 + 4 \iff 4x'^2 + y'^2 = 4$ where $x' = x-3$, $y' = y+2$, an ellipse with $C'(0,0)$, $V'(0,\pm 2)$, $M'(\pm 1,0)$. Thus $C(3,-2)$, $V(3,-2\pm 2)$, $M(3\pm 1,-2)$.

25. If (x_1,y_1) is the point of tangency, then the slope of the tangent line there is $y' = 4x_1 + 3$ and its equation is $y-y_1 = (4x_1+3)(x-x_1)$. Since $(2,-1)$ is on this line, its coordinates must satisfy the equation. Thus $-1-y_1 = (4x_1+3) \cdot (2-x_1)$ and since $y_1 = 2x_1^2 + 3x_1 + 1$ we obtain $-1 - 2x_1^2 - 3x_1 - 1 = 8x_1 - 4x_1^2 + 6 = 3x_1 \iff x_1^2 - 4x_1 - 4 = 0$ with solutions $x_1 = 2(1 \pm \sqrt{2})$.

28. If (x,y) is a point on the circle, then the cross section has the appearance shown. The area of this half ellipse is $\frac{1}{2}\pi cy = \frac{\pi}{2} c\sqrt{r^2-x^2}$. (See Example 4, Sec. 7.3.) Thus $V =$

$\dfrac{c\pi}{2} \displaystyle\int_{-r}^{r} \sqrt{r^2-x^2}\, dx$. The integral is the area of the

upper half of a circle of radius r, and has value $\frac{1}{2}\pi r^2$. Thus $V = \dfrac{c\pi}{2} \cdot \dfrac{\pi r^2}{2} = c(\pi r/2)^2$.

CHAPTER 8

EXPONENTIAL AND LOGARITHMIC FUNCTIONS

EXERCISES 8.1, page 367

1. $f'(x) = \dfrac{D_x(9x+4)}{9x+4} = \dfrac{9}{9x+4}$ by (8.6).

4. $f(x) = \ln(5x^2+1)^3 = 3\ln(5x^2+1)$ by (8.7,iii). Thus $f'(x) = 3\,\dfrac{D_x(5x^2+1)}{5x^2+1}$

$= \dfrac{3(10x)}{5x^2+1}$.

7. $f'(x) = \dfrac{D_x(3x^2-2x+1)}{(3x^2-2x+1)} = \dfrac{6x-2}{3x^2-2x+1}$.

10. $f(x) = \ln\sqrt{1-x^2} = \dfrac{1}{2}\ln(1-x^2) \implies f'(x) = (\dfrac{1}{2})\dfrac{D_x(1-x^2)}{1-x^2} = \dfrac{-x}{1-x^2}$.

13. $f(x) = \ln\sqrt{x} + \sqrt{\ln x} = (1/2)\ln x + (\ln x)^{1/2} \implies f'(x) = 1/2x + (1/2)(\ln x)^{-1/2} \cdot$
    $D_x(\ln x) = 1/2x + 1/2x\sqrt{\ln x}.$

16. $f(x) = \ln\sqrt{\dfrac{4+x^2}{4-x^2}} = \dfrac{1}{2}\ln\dfrac{4+x^2}{4-x^2} = \dfrac{1}{2}[\ln(4+x^2) - \ln(4-x^2)] \implies f'(x) =$

$\dfrac{1}{2}(\dfrac{2x}{4+x^2} - \dfrac{-2x}{4-x^2}) = \dfrac{8x}{16-x^4}$.

19. $f(x) = \ln\dfrac{\sqrt{x^2+1}}{(9x-4)^2} = \dfrac{1}{2}\ln(x^2+1) - 2\ln(9x-4) \implies f'(x) = (\dfrac{1}{2})2x/(x^2+1) -$
    $(2)9/(9x-4) = x/(x^2+1) - 18/(9x-4).$

22. $f'(x) = \dfrac{D_x\ln x}{\ln x} = \dfrac{1}{x\ln x}$.

25. $3y-x^2 + \ln xy = 2 \implies 3y' - 2x + \dfrac{D_x(xy)}{xy} = 0 \implies 3y' - 2x + \dfrac{xy'+y}{xy} =$

$3y' - 2x + \dfrac{y'}{y} + \dfrac{1}{x} = 0 \implies (3 + \dfrac{1}{y})y' = 2x - \dfrac{1}{x} \to y' = (\dfrac{2x^2-1}{x})\Big/(\dfrac{3y+1}{y})$.

28. $y^3 + x^2\ln y = 5x + 3 \implies 3y^2y' + (x^2D_x(\ln y) + \ln y\, D_x(x^2)) = 5 \to 3y^2y' +$
    $x^2(y'/y) + 2x\ln y = 5 \implies (3y^2 + x^2/y)y' = 5-2x\ln y \implies y' =$
    $(5-2x\ln y)/(3y^2+x^2/y).$

34. $x^3 - x\ln y + y^3 = 2x + 5 \implies 3x^2 - xy'/y - \ln y + 3y^2y' = 2.$ At $(2,1)$ this
    reduces to $12 - 2y' - \ln 1 + 3y' = 2$ or $y' = -10.$ Thus, equation:
    $(y-1) = -10(x-2).$

37. $v(t) = 2t - 4/(t+1)$, and $a(t) = 2 + 4/(t+1)^2$. Writing $v(t) =$
    $(2t(t+1) - 4)/(t+1) = 2(t-1)(t+2)/(t+1)$ we see that $v(t) < 0$ if $0 \le t < 1$;
    $v(t) > 0$ if $1 < t \le 4$. Thus the motion is to the left from $s(0) = 0$ to
    $s(1) = 1 - 4\ln 2$ from $t = 0$ to $t = 1$, then to the right to $s(4) = 16 - 4\ln 5$
    from $t = 1$ to $t = 4$.

40. (a)   (b)

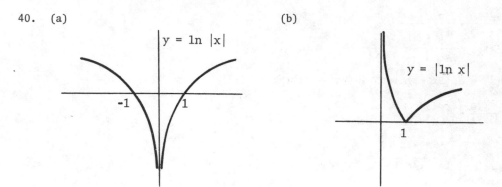

43. The derivative of the right side should equal the integrand on the left side.
     Verify this.

EXERCISES 8.2, page 374

1. $f'(x) = e^{-5x}D_x(-5x) = -5e^{-5x}$ by (8.17) with $g(x) = -5x$.

4. $f'(x) = e^{1-x^3}D_x(1-x^3) = -3x^2e^{1-x^3}$ by (8.17), $g(x) = 1-x^3$.

7. $f'(x) = e^{\sqrt{x+1}}D_x\sqrt{x+1} = (e^{\sqrt{x+1}})/2\sqrt{x+1}$.

10. $f'(x) = \frac{1}{2}(e^{2x}+2x)^{-1/2}D_x(e^{2x}+2x) = \frac{1}{2}(e^{2x}+2x)^{-1/2}(2e^{2x}+2) = (2e^{2x}+2)/2\sqrt{e^{2x}+2x}$

13. $f'(x) = 3(e^{4x}-5)^2D_x(e^{4x}-5) = 3(e^{4x}-5)^2(4e^{4x})$.

16. $f'(x) = D_x(e^{x^{1/2}} + e^{x/2}) = e^{\sqrt{x}}D_x(x^{1/2}) + e^{x/2}D_x(x/2) = e^{\sqrt{x}}/2\sqrt{x} + \frac{1}{2}e^{x/2}$.

19. $f'(x) = e^{-2x}D_x(\ln x) + \ln x\, D_x(e^{-2x}) = \dfrac{e^{-2x}}{x} + \ln x(-2)e^{-2x}$.

22. $f'(x) = \dfrac{\ln(e^x-1)D_x\ln(e^x+1) - \ln(e^x+1)D_x\ln(e^x-1)}{[\ln(e^x-1)]^2} =$

$\left\{\dfrac{[\ln(e^x-1)]e^x}{e^x + 1} - \dfrac{[\ln(e^x+1)]e^x}{e^x - 1}\right\}\Big/[\ln(e^x+1)]^2$.

25. $f'(x) = D_x(x) = 1$ for $x > 0$.   ($e^{\ln x} = x$ for $x > 0$ by (8.14))

28. $xe^y + 2x - \ln(y+1) = 3 \Longrightarrow xe^yy' + e^y + 2 - y'/(y+1) = 0 \Longrightarrow (xe^y - (y+1)^{-1})y'$
     $= -(2 + e^y) \Longrightarrow y' = (2+e^y)/((y+1)^{-1}-xe^y)$.

31. $y' = (x-1)D_xe^x + e^xD_x(x-1) + 3 D_x \ln x = (x-1)e^x + e^x + 3/x = xe^x + 3/x$.
     When $x = 1$, $y' = e+3$, and the tangent line is $y-2 = (e+3)(x-1)$.

34. $f(x) = \sqrt{e^x}$ exists for <u>all</u> x and, from $f'(x) =$
     $(1/2)\sqrt{e^x}$ and $f''(x) = (1/4)\sqrt{e^x}$, f is increasing
     and the graph is always CU. $g(x) = e^{\sqrt{x}}$ exists
     only for $x \geq 0$ and, from $g'(x) = e^{\sqrt{x}}/2\sqrt{x}$ and
     $g''(x) = (\sqrt{x}-1)e^{\sqrt{x}}/4x^{3/2}$, g is increasing and
     the graph is CD on $(0,1)$, CU on $(1,\infty)$.

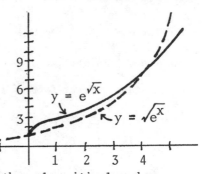

37. $f'(x) = xe^x + e^x = (x+1)e^x = 0$ only at $x = -1$, the only critical number.
     Since $e^x > 0$ for all x, $x < -1 \implies x+1 < 0 \implies f'(x) < 0$ and f is decreasing
     on $(-\infty,-1]$. Similarly f is increasing on $[-1,\infty)$, and $f(-1) = -e^{-1}$ is a local
     minimum. $f''(x) = (x+2)e^x$ changes from negative to positive at $x = -2$. Thus
     the graph is CD on $(-\infty,-2)$, CU on $(-2,\infty)$ and has a PI at $x = -2$.

40. $f'(x) = (1-x)e^{-x}$ which is 0 at $x = 1$, > 0 if $x < -1$ and < 0 if $x > -1$. Thus
     f is increasing on $(-\infty,-1]$, decreasing on $[-1,\infty)$, and $f(1) = e^{-1}$ is a local
     minimum. $f''(x) = (x-2)e^{-x}$ changes from negative to positive at $x = 2$. Thus
     the graph is CD on $(-\infty,2)$, CU on $(2,\infty)$ and has a PI at $x = 2$.

NOTE. In #43-46, use is made of the formulas $y = e^{cx} \implies y' = ce^{cx} \implies y'' = c^2e^{cx} \implies$
       $y''' = c^3e^{cx}$ and the fact that $e^{cx} \neq 0$ for all x.

43. Substituting $y = e^{cx}$, $y'' - 3y' + 2y = c^2e^{cx} - 3ce^{cx} + 2e^{cx} = (c^2-3c+2)e^{cx} = 0$
     if $c^2 - 3c + 2 = 0$. Factoring, $c^2 - 3c + 2 = (c-2)(c-1) = 0$ if $c = 1$ and $c = 2$.

46. $y''' - y'' - 6y' = c^3e^{cx} - c^2e^{cx} - 6ce^{cx} = (c^3 - c^2 - 6c)e^{cx} = 0$ if $c^3 - c^2 - 6c$
     $= 0$. Factoring: $c^3 - c^2 - 6c = c(c^2 - c - 6) = c(c-3)(c+2) = 0$ if $c = 0$,
     3 and -2.

EXERCISES 8.3, page 381

1. Let $u = x^2+1$. Then $du = 2x\,dx$, $x\,dx = (1/2)\,du$. $\int \frac{x}{x^2+1}\,dx = \frac{1}{2}\int \frac{du}{u} = \frac{1}{2}\ln|u|$

   $+ C = \frac{1}{2}\ln|x^2+1| + C = \frac{1}{2}\ln(x^2+1) + C$ since $x^2+1 > 0$. ($\ln \sqrt{x^2+1} + C$ is

   another correct form of the answer.)

4. Let $u = x^4-5$, $du = 4x^3dx$, $x^3dx = (1/4)\,du$. $\int \frac{x^3}{x^4-5}\,dx = \frac{1}{4}\int \frac{du}{u} = \frac{1}{4}\ln|u| + C =$

   $\frac{1}{4}\ln|x^4-5| + C$.

7. With $u = x^3+1$, $du = 3x^2dx$, $\int \frac{x^2}{x^3+1}\,dx = \frac{1}{3}\int \frac{du}{u} = (1/3)\ln|u| + C = (1/3)\ln|x^3+1|+C$.

10. Let $u = 4-5x$, $du = -5dx$. When $x = -1$, $u = 9$ and when $x = 0$, $u = 4$. Thus

    $\int_{-1}^{0} \frac{dx}{4-5x} = -\frac{1}{5}\int_{9}^{4} \frac{du}{u} = -\frac{1}{5}\ln|u|\Big]_{9}^{4} = -\frac{1}{5}(\ln 4 - \ln 9) = \frac{1}{5}\ln(9/4)$.

13. $\int (x + e^{5x})\,dx = \frac{x^2}{2} + \int e^{5x}dx$. Let $u = 5x$, $du = 5\,dx$, $dx = (1/5)\,du$. Using

    this in the remaining integral, we get $\frac{x^2}{2} + \frac{1}{5}\int e^u du = \frac{x^2}{2} + \frac{e^u}{5} + C = \frac{x^2}{2} + \frac{e^{5x}}{5}+C$.

16. With $u = \ln x$, $du = \frac{1}{x}dx$, $\int \frac{1}{x(\ln x)^2}\,dx = \int \frac{1}{u^2}\,du = -\frac{1}{u} + C = -\frac{1}{\ln x} + C$.

19. With $u = \sqrt{x}$, $du = \frac{1}{2\sqrt{x}}\,dx$, $\int \frac{e^{\sqrt{x}}}{\sqrt{x}}\,dx = 2\int e^u du = 2e^u + C = 2e^{\sqrt{x}} + C$.

22. With $u = e^x+1$, $du = e^x dx$, $\int \frac{e^x}{(e^x+1)^2}\,dx = \int \frac{1}{u^2}\,du = -\frac{1}{u} + C = -1/(e^x+1) + C$.

25. Since $x^2 + 2x + 1 = (x+1)^2$, we let $u = (x+1)$, $du = dx$ so that $\int \frac{1}{x^2+2x+1}\,dx$

    $= \int \frac{1}{(x+1)^2}dx = \int \frac{1}{u^2}\,du = \int u^{-2}du = -u^{-1} + C = -1/(x+1) + C$.

28. $\int \frac{x^2+3x + 1}{x}\,dx = \int (x + 3 + \frac{1}{x})\,dx = \frac{x^2}{2} + 3x + \ln|x| + C$.

31. $A = \int_{1}^{2} (\frac{1}{x} - e^{-x})\,dx = \ln|x| + e^{-x}\Big]_{1}^{2} = (\ln 2 + e^{-2}) - (\ln 1 + e^{-1}) = \ln 2 +$

    $e^{-2} - e^{-1} \approx 0.460$.   (Recall that $\ln 1 = 0$.)

34. $V = \pi\int_{1}^{4} (\frac{1}{\sqrt{x}})^2 dx = \pi\int_{1}^{4} \frac{1}{x}dx = \pi \ln x\Big]_{1}^{4} = \pi \ln 4 \approx 1.39\pi$.

37. (a) $f'(x) = e^x \Longrightarrow L_0^1 = \int_{0}^{1} \sqrt{1+(e^x)^2}\,dx$

    (b) With $a = 0$, $b = 1$, $n = 5$, the partitioning points are 0, .2, .4, .6, .8

and 1. With $g(x) = \sqrt{1+(e^x)^2}$ we obtain $L_0^1 \approx \frac{(1-0)}{2(5)}[g(0) + 2g(.2) + 2g(.4) +$

$2g(.6) + 2g(.8) + g(1)]$.

(c) $L_0^1 \approx (0.1)[(1.414 + 2(1.579) + 2(1.796) + 2(2.078) + 2(2.440) + 2(2.896)]$

$= 2.010$.

40. Following the outline of steps on p. 380, we obtain: $y = \sqrt{4x+7}(x-5)^3 =$

$(4x+7)^{1/2}(x-5)^3 \implies \ln y = (1/2)\ln(4x+7) + 3 \ln (x-5) \implies \frac{y'}{y} = \frac{1}{2}\frac{4}{4x+7} +$

$3\frac{1}{x-5} = \frac{2x - 10 + 12x + 21}{(4x+7)(x-5)} = \frac{14x+11}{(4x+7)(x-5)}$. Thus $y' = \frac{14x+11}{(4x+7)(x-5)} y =$

$\frac{14x+11}{(4x+7)(x-5)} \sqrt{4x+7}(x-5)^3 = \frac{(14x+11)(x-5)^2}{\sqrt{4x+7}}$.

43. $\ln y = \frac{1}{3}\ln(2x+1)+2 \ln(4x-1)+4 \ln(3x+5)$. $\frac{y'}{y} = \frac{2}{3(2x+1)} + \frac{2(4)}{4x-1} + \frac{4(3)}{3x+5}$. Multiply-

ing by y and substituting for y yields the indicated answer in the text.

46. $\ln y = \frac{2}{3} \ln(x^2+3) + 4 \ln(3x-4) - \frac{1}{2} \ln x$. $\frac{y'}{y} = \frac{2(2x)}{3(x^2+3)} + \frac{4(3)}{3x-4} - \frac{1}{2x}$.

$y' = [\frac{4x}{3(x^2+3)} + \frac{12}{3x-4} - \frac{1}{2x}]\frac{(x^2+3)^{2/3}(3x-4)^4}{\sqrt{x}}$.

49. $f(x) = \ln|1-e^{-2x}|^3 = 3 \ln|1-e^{-2x}|$. $f'(x) = \frac{3D_x(1-e^{-2x})}{1-e^{-2x}} = \frac{3(2e^{-2x})}{1-e^{-2x}}$.

EXERCISES 8.4, page 389

1. $f'(x) = 7^x \ln 7$ immediately from (8.25).

4. $f'(x) = 9^{\sqrt{x}} \ln 9 \, D_x\sqrt{x} = (9^{\sqrt{x}} \ln 9)/2\sqrt{x}$ by (8.26).

7. $f'(x) = 5^{3x-4}\ln 5 \, D_x(3x-4) = 5^{3x-4}(\ln 5) \cdot 3$.

10. $f'(x) = 10(10^x + 10^{-x})^9 D_x(10^x + 10^{-x}) = 10(10^x + 10^{-x})^9(10^x\ln 10 - 10^{-x}\ln 10)$.

13. $f'(x) = D_x5 \log_{10}(3x^2 + 2) = \frac{5(6x)}{(\ln 10)(3x^2+2)}$.

16. $f'(x) = D_x(\log_{10}|1-x^2| - \log_{10}|2-5x^3|) = \frac{(-2x)}{(\ln 10)(1-x^2)} - \frac{(-15x^2)}{(\ln 10)(2-5x^3)}$.

19. $f'(x) = D_x(x^e) + D_x(e^x) = ex^{e-1} + e^x$ since the first function is a <u>power</u>
    <u>function</u>; the second is the familiar exponential.

22. Method 1. $D_x(x^{x^2+4}) = D_xe^{(x^2+4)\ln x} = e^{(x^2+4)\ln x}D_x(x^2+4)\ln x =$

$x^{x^2+4}[(x^2+4)/x + 2x \ln x]$.

Method 2. $y = x^{x^2+4} \iff \ln y = (x^2+4) \ln x \implies \dfrac{y'}{y} = \dfrac{x^2+4}{x} + 2x \ln x \implies$

$y' = y((x^2+4)/x + 2x \ln x)$.

25. Let $u = -x^2$, $du = -2x\, dx$, $x\, dx = -(1/2)du$. Then $\displaystyle\int x3^{-x^2}\, dx = -\dfrac{1}{2}\int 3^u du =$

$-\dfrac{1}{2}\dfrac{3^u}{\ln 3} + C = -\dfrac{3^{-x^2}}{2 \ln 3} + C.$

28. Let $u = 3^x+4$, $du = 3^x \ln 3\, dx$. Then $3^x\, dx = (1/\ln 3)du$ and $\displaystyle\int \dfrac{3^x}{\sqrt{3^x+4}}\, dx =$

$\dfrac{1}{\ln 3}\displaystyle\int \dfrac{du}{\sqrt{u}} = \dfrac{1}{\ln 3}\int u^{-1/2}du = \dfrac{2}{\ln 3}u^{1/2} + C = \dfrac{2}{\ln 3}\sqrt{3^x+4} + C.$

31. With $u = x^3$, $du = 3x^2 dx$, $\displaystyle\int x^2 2^{x^3}\, dx = \dfrac{1}{3}\int 2^u du = 2^u/3 \ln 2 + C = 2^{x^3}/3 \ln 2 + C.$

34. With $u = 10^x - 10^{-x}$, $du = (10^x + 10^{-x})\ln 10\, dx$, $\displaystyle\int \dfrac{10^x + 10^{-x}}{10^x - 10^{-x}}\, dx =$

$\dfrac{1}{\ln 10}\displaystyle\int \dfrac{du}{u} = \dfrac{1}{\ln 10}\ln|u| + C = (\ln|10^x - 10^{-x}|)/\ln 10 + C.$

37. $y = x3^x \implies y' = x3^x\ln 3 + 3^x$. When $x = 1$, $y' = 3 \ln 3 + 3$ and the tangent

line is $(y-3) = 3(1 + \ln 3)(x-1)$; the normal line is $(y-3) = \dfrac{-1}{3(1 + \ln 3)}(x-1).$

40. HINT: Use $\log_a x = \ln x/\ln a$ on (8.7). For example, $\log_a pq = \ln pq/\ln a =$

$(\ln p + \ln q)/\ln a = \ln p/\ln a + \ln q/\ln a = \log_a p + \log_a q.$

EXERCISES 8.5, page 396

1. The procedure for finding $q(t)$ is exactly the same as in Example 1. We ob-

tain $q(t) = (5000)3^{t/10}$. After 20 hours, $t = 20$, and $q(20) = (5000)3^2 =$

45,000. To find when $q(t)$ is 50,000 we solve $(5000)3^{t/10} = 50{,}000 \iff$

$3^{t/10} = 10 \iff (t/10)\ln 3 = \ln 10 \iff t = (10 \ln 10)/\ln 3 \approx 23/1.1 \approx$

21 hours.

4. If $P(t) = $ the population t years from the present, then $\dfrac{dP}{dt} = .05P$ and $P(t) =$

$P(0)e^{.05t} = 500{,}000\, e^{.05t}$. In 10 years $t = 10$ and $p(10) = 500{,}000e^{.5} \approx$

$500{,}000(1.65) = 825{,}000.$

7. Let $y(t)$ be the temperature of the thermometer t minutes after it's brought

in. (Assume that the reading is the same as the actual temperature.) Then

$\dfrac{dy}{dt} = c(y-70)$, and $y(10) = 40$ which yields $\dfrac{dy}{y-70} = c\, dt \implies \ln|y-70| = ct + b$

$\implies |y-70| = ke^{ct}$. Since the instrument is cooler than 70, $y-70 < 0$ and the last formula becomes $70-y = ke^{ct}$. Setting $t = 0$, $y(0) = 40$ and $k = 30$. Using $y(5) = 60$ we obtain $10 = 30e^{5c}$ or $c = (-1/5)\ln 3$. To find when it registers 65, we solve $70-65 = 30e^{(-1/5)(\ln 3)t} \iff (-1/5)(\ln 3)t = \ln(1/6) = -\ln 6 \implies t = (5 \ln 6)/\ln 3 \approx 5(1.792)/1.099 \approx 8.2$ minutes.

10. The data given yields $\frac{dy}{dt} = c(k-y)$, $y(0) = 0$ where c is the proportionality constant. (The initial value is 0 since at the start all the sugar is de-composed.) Solving, as in Example 3 or Example 7, $\frac{dy}{k-y} = c\ dt \implies -\ln(k-y) = ct + b$ (since $k > y$). Setting $t = 0$, we get $-\ln k = b$. Thus $\ln(k-y) = -ct + \ln k \implies k-y - ke^{-ct} \implies y = k(1 - e^{-ct})$.

13. (a) Writing $G'(t) = ABke^{-Bt}e^{-Ae^{-Bt}}$, we see that if $t = (\ln A)/B$ then $e^{Bt} = e^{\ln A} = A$, $e^{-Bt} = 1/A$, and $G'((\ln A)/B) = ABk(1/A)e^{-A/A} = Bke^{-1}$. That this is a maximum for G' follows from the first derivative test. All factors in $G''(t)$ are positive except $(-1 + Ae^{-Bt})$ which is < 0 if $t > (\ln A)/B$ and > 0 if $t < (\ln A)/B$.

(b) $B > 0 \implies e^{-Bt} \to 0$ as $t \to \infty$ and $e^{-Ae^{-Bt}} \to e^0 = 1$. Thus $\lim_{t\to\infty} G'(t) = ABk(0)1 = 0$.

(c) $G(t) = ke^{-Ae^{-Bt}} \to ke^{-(A)(0)} = k$ as $t \to \infty$.

16. First note that $f(0) = a$ and $f(x) \to a+b$ as $x \to \infty$, $f(x) \to -\infty$ as $x \to -\infty$. $f'(x) = bce^{-cx} > 0 \implies f$ is increasing. $f''(x) = -c^2be^{-cx} < 0 \implies$ the graph is CD on $(-\infty,\infty)$.

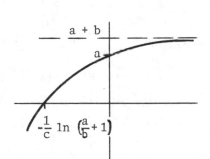

19. Let $V(t)$ be the value t years after purchase. We are given that $V(t) = Ae^{kt}$ where $k < 0$. $V(0) = 20,000 \implies A = 20,000$ so that $V(t) = 20,000\ e^{kt}$. $V(2) = 16,000 \implies 20,000\ e^{2k} = 16,000 \implies e^{2k} = 0.8 \implies e^k = \sqrt{0.8}$. Thus $V(t) = 20,000(e^k)^t = 20,000\sqrt{0.8}^t = 20,000(0.8)^{t/2}$. Then after one more year, $t = 3$ and $V(3) = 20,000(0.8)^{3/2} \approx 14,311$.

EXERCISES 8.6, page 402

1. $f'(x) = 1/\sqrt{2x+3} > 0 \Longrightarrow$ f is increasing on [1,11]
    and has an inverse with domain [f(1),f(11)] =
    $[\sqrt{5},5]$. As in Sec. 1.6, we find f^{-1} by solving y =
    $\sqrt{2x+3}$ to obtain $y^2 = 2x+3$ and x = $(y^2-3)/2$. Thus
    $f^{-1}(x) = (x^2-3)/2$. Directly, $D_x f^{-1}(x) = x$. Using
    (8.42) with g = f^{-1}, $D_x f^{-1}(x) = 1/f'(f^{-1}(x)) =$
    $\sqrt{2f^{-1}(x) + 3}$ (from the formula for f' in the first
    line) = $\sqrt{(x^2-3) + 3} = x$ since x > 0.

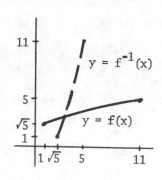

4. $f'(x) = 2x-4 < 0 \Longrightarrow$ f is decreasing on [-1,1]
    and has an inverse with domain [f(1),f(-1)] =
    [2,10]. Setting y = x^2-4x+5 and solving
    for x, we get x = $2 - \sqrt{y-1}$ by the quadratic
    formula. (The negative sign must be se-
    lected for x to lie in [-1,1].) Thus $f^{-1}(x)$
    = $2 - \sqrt{x-1}$. Using (8.42), $D_x f^{-1}(x) =$
    $1/(2f^{-1}(x) - 4) = -1/2\sqrt{x-1}$, the same as
    computing it directly.

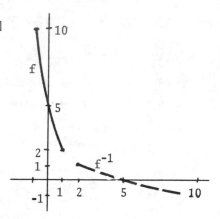

7. $f'(x) = -2xe^{-x^2} < 0$ if x > 0. Thus f is decreas-
    ing on [0,∞) and has an inverse with domain equal
    to the range of f. Since f(0) = 1 and f(x) → 0
    as x → ∞, f^{-1} has domain (0,1]. Solving y =
    e^{-x^2} we get $-x^2 = \ln y \Longrightarrow x = \sqrt{-\ln y}$. Thus
    $f^{-1}(x) = \sqrt{-\ln x}$ and $D_x f^{-1}(x) = -1/2x\sqrt{-\ln x}$
    by either method.

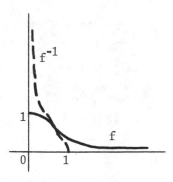

10. $f'(x) = e^x - e^{-x} > 0$ if x > 0. Thus f is in-
    creasing on [0,∞) and has an inverse with do-
    main equal to the range of f. Since f(0) = 2
    and f(x) → ∞ as x → ∞, the domain of f^{-1} is (2,∞).
    Next,   y = $e^x + e^{-x} \Longrightarrow e^x - y + e^{-x} = 0$
    $\Longrightarrow e^{2x} - ye^x + 1 = 0$. Now solve for e^x by
    the quadratic formula! $e^x = (y \pm \sqrt{y^2-4})/2$.

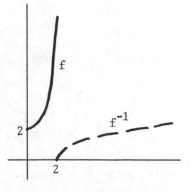

The + sign must be chosen so that $x = \ln[(y + \sqrt{y^2-4})/2]$. Thus $f^{-1}(x) = \ln[(x + \sqrt{x^2-4})/2]$ and $Df^{-1}(x) = [1 + x/\sqrt{x^2-4}]/(x + \sqrt{x^2-4}) = 1/\sqrt{x^2-4}$.

(NOTE: The quadratic formula can be used since $e^{2x} = (e^x)^2$. Also, if the $-$ sign had been selected, we would have $x = \ln[(y - \sqrt{y^2-4})/2]$. But as $y \to \infty$, $y - \sqrt{y^2-4} \to 0$ and x would tend to $-\infty$ contrary to the earlier observation that $y = f(x)$ and x go to ∞ simultaneously.)

13. $f'(x) = 4e^{2x}/(e^{2x}+1)^2 > 0 \Longrightarrow f$ has an inverse. Using (8.42), if g is the inverse function, $g'(0) = 1/f'(g(0)) = 1/f'(0) = 1$.

16. f is decreasing on $(-\infty,2]$, increasing on $[2,\infty)$ and, thus, has no inverse. On any subset of either of these intervals, f will have an inverse.

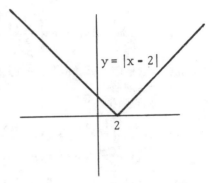

EXERCISES 8.7 (Review), page 403

1. $f'(x) = (1-2x)D_x\ln|1-2x| + \ln|1-2x|\cdot D_x(1-2x) = \dfrac{(1-2x)(-2)}{(1-2x)} + (-2)\ln|1-2x| = -2(1 + \ln)|1-2x|)$.

4. $f(x) = \log_{10}|2-9x| - \log_{10}|1-x^2| \Longrightarrow f'(x) = \dfrac{1}{\ln 10}(\dfrac{-9}{2-9x} - \dfrac{-2x}{1-x^2})$.

7. By (8.14), $f(x) = e^{\ln(x^2+1)} = x^2+1$, and $f'(x) = 2x$.

10. $f(x) = \dfrac{1}{4}\ln\dfrac{x}{3x+5} = \dfrac{1}{4}(\ln x - \ln(3x+5)) \Longrightarrow f'(x) = \dfrac{1}{4}(\dfrac{1}{x} - \dfrac{3}{3x+5}) = \dfrac{5}{4x(3x+5)}$.

13. $f'(x) = x^2 e^{1-x^2}(-2x) + 2xe^{1-x^2} = 2x(1-x^2)e^{1-x^2}$.

16. $f'(x) = 5^{3x}(\ln 5)(3) + 5(3x)^4 \cdot 3$.

19. $f'(x) = 10^{\ln x}(\ln 10)D_x\ln x = 10^{\ln x}(\ln 10)/x$.

22. $\ln(x+y) + x^2 - 2y^3 = 1 \Longrightarrow (1+y')/(x+y) + 2x - 6y^2y' = 0 \Longrightarrow y' = (2x + \dfrac{1}{x+y})\Big/(6y^2 - \dfrac{1}{x+y})$.

25. $a(t) = e^{t/2} \Longrightarrow v(t) = 2e^{t/2} + C$. $v(0) = 6 \Longrightarrow C = 4$ and $v(t) = 2e^{t/2} + 4$. Since $v(t) > 0$, the distance travelled from $t = 0$ to $t = 4$ is $s(4) - s(0) =$

$$\int_0^4 v(t)\ dt = 4e^{t/2} + 4t\ \Big]_0^4 = 4e^2 + 12 \approx 41.56 \text{ cm.}$$

28. With $u = \ln x$, $du = (1/x)dx$, $\displaystyle\int \frac{dx}{x \ln x} = \int \frac{1}{u}\,du = \ln|u| + C = \ln|\ln x| + C$.

31. By long division $\displaystyle\int \frac{x^2}{3x+2}\,dx = \int [\frac{x}{3} - \frac{2}{9} + (\frac{4}{9})\frac{1}{3x+2}]\,dx = \frac{x^2}{6} - \frac{2x}{9} + \frac{4}{27}\ln|3x+2| + C$.

34. $\displaystyle\int \frac{(e^{2x}+e^{3x})^2}{e^{5x}}\,dx = \int \frac{(e^{2x})^2 + 2(e^{2x})(e^{3x}) + (e^{3x})^2}{e^{5x}}\,dx = \int \frac{e^{4x} + 2e^{5x} + e^{6x}}{e^{5x}}\,dx =$

$\displaystyle\int (e^{-x}+2+e^{x})dx = -e^{-x} + 2x + e^{x} + C$.

37. By long division $\displaystyle\int \frac{x^2+1}{x+1}\,du = \int (x - 1 + \frac{2}{x+1})\,dx = \frac{x^2}{2} - x + 2\ln|x+1| + C$.

40. With $u = x^2$, $du = 2x\,dx$, $\displaystyle\int x\,10^{x^2}\,dx = \frac{1}{2}\int 10^u\,du = \frac{1}{2\ln 10}\,10^u + C =$
    $\displaystyle\frac{10^{x^2}}{2\ln 10} + C$.

43. $y' = xe^{1/x^3}(-3/x^4) + e^{1/x^3} - 2x/(2-x^2)$. When $x = 1$, $y' = -3e + e - 2 = -2(1+e)$. Thus, the tangent line is $y-e = -2(1+e)(x-1)$.

46. $f'(x) = 6x^2-8 < 0$ on $[-1,1]$. Thus f is decreasing and has an inverse g. Since $f(0) = 5$, $g(5) = 0$ and $g'(5) = 1/f'(g(5)) = 1/f'(0) = -1/8$.

49. Let $y = y(t)$ be the amount (in lbs) dissolved t hours after 1:00 PM. Then $10-y$ remain and $\dfrac{dy}{dt} = k(10-y) \Longrightarrow \dfrac{dy}{10-y} = k\,dt \Longrightarrow \ln(10-y) = -kt - C$. The conditions to be used are $y(0) = 0$ and $y(3) = 5$. (At 4 PM, $t = 3$, and half is dissolved.) These conditions imply $C = -\ln 10$, $k = (\ln 2)/3$ and thus $y = 10(1 - e^{-(\ln 2)t/3}) = 10(1 - (\frac{1}{2})^{t/3})$.

(a) 2 more lbs will be dissolved (after the 5 that were dissolved at 4 PM) when $y = 7$. Set $y = 7$ and solve for t: $1 - (\frac{1}{2})^{t/3} = .7 \Longrightarrow (1/2)^{t/3} = .3 \Longrightarrow (t/3)\ln(.5) = \ln(.3) \Longrightarrow t = 3\ln(.3)/\ln(.5) \approx 3(-1.204)/(-0.693) \approx 5.21$ (i.e. at 6.21 PM or $\approx$ 6:14 PM).

(b) At 8 PM, $t = 7$ and $y = 10(1 - (1/2)^{7/3}) \approx 10(1 - .1984) = 8.016$ lbs.

CHAPTER 9

OTHER TRANSCENDENTAL FUNCTIONS

<u>EXERCISES 9.1, page 411</u>

1. $\lim\limits_{t \to 0} \dfrac{\cos t}{1 - \sin t} = \dfrac{1}{1-0} = 1$.

4. $\lim\limits_{x \to 0} \dfrac{\sin(x/2)}{x} = \lim\limits_{x \to 0} (\dfrac{1}{2}) \dfrac{\sin(x/2)}{(x/2)} = \dfrac{1}{2} \cdot 1 = \dfrac{1}{2}$.

7. $\lim\limits_{x \to 0} x \cot x = \lim\limits_{x \to 0} \dfrac{x \cos x}{\sin x} = \lim\limits_{x \to 0} \dfrac{x}{\sin x} \cos x = 1 \cdot 1 = 1$.

10. Let $t = x-\pi$. Then $x \to \pi \iff t \to 0$. $\lim\limits_{x \to \pi} \dfrac{\sin x}{x-\pi} = \lim\limits_{t \to 0} \dfrac{\sin(t+\pi)}{t} = \lim\limits_{t \to 0} \dfrac{-\sin t}{t}$
    $= -1$.

13. $\lim\limits_{x \to 3} \dfrac{x^2-6x+9}{\sin^2(x-3)} = \lim\limits_{x \to 3} [\dfrac{x-3}{\sin(x-3)}]^2$. With $t = x-3$, as $x \to 3$ then $t \to 0$, and the

    desired limit is the same as $\lim\limits_{t \to 0} [\dfrac{t}{\sin t}]^2 = \lim\limits_{t \to 0} \dfrac{1}{(\sin t/t)^2} = 1$ by (9.3).

16. $\lim\limits_{\alpha \to 0} \alpha^2 \csc^2\alpha = \lim\limits_{\alpha \to 0} \dfrac{\alpha^2}{\sin^2\alpha} = \lim\limits_{\alpha \to 0} (\dfrac{\alpha}{\sin \alpha})^2 = 1$.

19. $\lim\limits_{x \to \infty} x \cos \dfrac{1}{x} = \infty$ since as $x \to \infty$, $\dfrac{1}{x} \to 0$ and $\cos(\dfrac{1}{x}) \to 1$. Thus the given function

    is the product of x, increasing without bound, and $\cos(\dfrac{1}{x})$, nearly 1.

22. HINT. Multiply and divide by a, let $t = ax$ and use (9.5).

<u>EXERCISES 9.2, page 419</u>

1. $D_x \cot u = D_x \dfrac{\cos u}{\sin u} = \dfrac{\sin u\, D_x \cos u - \cos u\, D_x \sin u}{\sin^2 u} = \dfrac{-\sin^2 u\, D_x u - \cos^2 u\, D_x u}{\sin^2 u}$

    $= \dfrac{-D_x u}{\sin^2 u} = -\csc^2 u\, D_x u$. (9.12) is derived similarly. Recall: $\sin^2 u + \cos^2 u = 1$.

4. By (9.8), $f'(x) = -\sin(x/2)\, D_x(x/2) = -(1/2)\sin(x/2)$.

7. By (9.11), $f'(x) = -\csc^2(x^3-2x)\, D_x(x^3-2x) = -[\csc^2(x^3-2x)](3x^2-2)$, or, better,
    $-(3x^2-2)\csc^2(x^3-2x)$.

10. $f'(x) = 3(\tan^2 6x)D_x \tan 6x = 3 \tan^2 6x \sec^2 6x\, D_x(6x) = 18 \tan^2 6x \sec^2 6x$.

13. $f'(x) = x^2\, D_x \csc 5x + (D_x x^2)\csc 5x = -x^2 \csc 5x \cot 5x\, D_x(5x) + 2x \csc 5x$
    $= -5x^2 \csc 5x \cot 5x + 2x \csc 5x$.

16. $f'(x) = x^2 \cdot 3 \sec^2 4x \cdot D_x \sec 4x + D_x(x^2)\sec^3 4x = 3x^2 \sec^2 4x(\sec 4x \tan 4x) \cdot$
    $D_x(4x) + 2x \sec^3 4x = 12x^2 \sec^3 4x \tan 4x + 2x \sec^3 4x$.

19. $f'(x) = 3 \cot^2(3x+1) D_x \cot(3x+1) = 3 \cot^2(3x+1)(-\csc^2(3x+1)) \cdot 3.$

22. $f'(x) = \dfrac{[(\tan 2x + 1)(\sec 2x \tan 2x)(2) - \sec 2x(\sec^2 2x)(2)]}{(\tan 2x + 1)^2}.$

25. $f'(x) = e^{-3x}D_x \tan \sqrt{x} + \tan \sqrt{x} D_x e^{-3x} = (e^{-3x} \sec^2\sqrt{x})/2\sqrt{x} - 3(\tan \sqrt{x})e^{-3x}.$

28. $f'(x) = -\csc(\cot 4x)\cot(\cot 4x)D_x \cot 4x$

$= (\csc(\cot 4x))(\cot(\cot 4x))(\csc^2 4x)(4).$

31. $f'(x) = [(x^3+1)D_x \csc 3x - \csc 3x D_x(x^3+1)]/(x^3+1)^2$

$= [-(x^3+1)\csc 3x \cot 3x D_x(3x) - 3x^2 \csc 3x]/(x^3+1)^2$

$= -3[(x^3+1)\csc 3x \cot 3x + x^2\csc 3x]/(x^3+1)^2.$

34. Using logarithmic differentiation, $y = (\tan x)^{3x} \Longrightarrow \ln y = 3x \ln(\tan x) \Longrightarrow$

$\dfrac{y'}{y} = 3x \dfrac{D_x(\tan x)}{\tan x} + 3 \ln(\tan x) = 3x \dfrac{\sec^2 x}{\tan x} + 3 \ln(\tan x) \Longrightarrow y' =$

$(3x \dfrac{\sec^2 x}{\tan x} + 3 \ln(\tan x))(\tan x)^{3x}.$

37. $\dfrac{dy}{dx} = \cos x - x(-\sin x) - \cos x = x \sin x$

$\dfrac{d^2y}{dx^2} = x \cos x + \sin x.$

40. $\dfrac{dy}{dx} = \dfrac{(\cos x + 1)(-\sin x) - (\cos x - 1)(-\sin x)}{(\cos x + 1)^2} = (-2 \sin x)/(\cos x + 1)^2.$

$\dfrac{d^2y}{dx^2} = \dfrac{(\cos x + 1)^2(-2 \cos x) + 2 \sin x(2)(\cos x + 1)(-\sin x)}{(\cos x + 1)^4}$

$= (-2 \cos x(\cos x + 1) - 4 \sin^2 x)/(\cos x + 1)^3.$

43. $e^x\cos y = xe^y \Longrightarrow e^x(-\sin y)y' + e^x\cos y = xe^yy' + e^y \Longrightarrow (e^x\sin y + xe^y)y'$

$= e^x\cos y - e^y,$ which yields the answer given.

46. $f'(x) = -\sin x - \cos x = 0$ when $\sin x =$
$-\cos x \Longrightarrow \tan x = -1 \Longrightarrow x = 3\pi/4, 7\pi/4,$
$11\pi/4, 15\pi/4.$ $f''(x) = -\cos x + \sin x.$
$f''(3\pi/4) = f''(11\pi/4) = 2/\sqrt{2} > 0,$ $f''(7\pi/4)$
$= f''(15\pi/4) = -2/\sqrt{2} < 0.$ Thus local minima
$f(3\pi/4) = f(11\pi/4) = -\sqrt{2}$ and local maxima
$f(7\pi/4) = f(15\pi/4) = \sqrt{2}.$

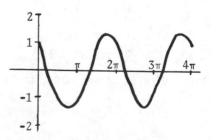

49. $y' = 24 \sin^2 x \cos x.$ At $x = \pi/6,$ $y' = 24 \cdot (1/2)^2 \cdot (\sqrt{3}/2) = 3\sqrt{3}.$ Thus the
tangent line: $(y-1) = 3\sqrt{3}(x-\pi/6)$ and the normal line: $(y-1) = (-1/3\sqrt{3})(x-\pi/6).$

52. With $f(x) = \cot x,$ $x = 45^\circ = \pi/4$ rad., $\Delta x = 1^\circ = \pi/180$ rad. $\Delta y \approx dy =$
$f'(\pi/4)(\pi/180) = -(\csc^2(\pi/4))(\pi/180) = -2\pi/180 \approx -0.035.$

55. $v(t) = s'(t) = -6 \sin 2t$, $a(t) = v'(t) = -12 \cos 2t$. The motion described in
     the answer is immediate from the periodicity of $s(t) = 3 \cos 2t$.

57. HINT: Let R be the radius of the circular sheet and θ,
     the central angle of the section cut out. Then the
     circumference of the remainder of the sheet is
     $(2\pi-\theta)R$ which is also the circumference of the top
     of the cup. If r is the radius of the top, h the
     height, and V the volume of the cup, then $2\pi r =$
     $(2\pi-\theta)R$, $h^2 + r^2 = R^2$, $V = \pi r^2 h/3$. Get V in
     terms of θ and differentiate, or use an implicit
     method.

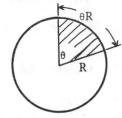

58. Strangely, this maximum problem is solved by con-
     sidering a minimum problem. Let $L = L_1 + L_2$ be,
     as shown, the distance between points on the
     walls measured on a line touching the inner
     corner. As $\theta \to 0$ or $\theta \to \pi/2$, $L \to \infty$, and it
     is clear that there is an angle θ_0 which makes
     L a minimum. Let L_0 be the minimum value of L.
     A rod of length L_0 <u>will</u> just fit around the
     corner. It will touch both walls when the
     turning angle is θ_0, but there is excess room
     for any other angle. However, any rod of
     length $> L_0$ will not fit around the corner. It will be jammed tight touching
     the corner and the walls, at some angle $\theta \neq \theta_0$, and further rotation will be
     impossible. So, our problem is to minimize L as a function of θ. By elemen-
     tary trigonometry, $L = L_1 + L_2 = \dfrac{3}{\cos \theta} + \dfrac{4}{\sin \theta}$. $\dfrac{dL}{d\theta} = \dfrac{3 \sin \theta}{\cos^2\theta} - \dfrac{4 \cos \theta}{\sin^2\theta} =$
     $\dfrac{3 \sin^3\theta - 4 \cos^3\theta}{\cos^2\theta \sin^2\theta} = 0$ when $3 \sin^3\theta = 4 \cos^3\theta$ or $\tan^3\theta = \dfrac{4}{3} \Longrightarrow \tan \theta = \sqrt[3]{4/3}$.
     Using tables or a calculator, we find that $\theta \approx 47^\circ 45'$ and $L_0 \approx 9.87$ ft.

EXERCISES 9.3, page 425

1. With $u = 4x$, $\displaystyle\int \sin 4x\, dx = (1/4) \int \sin u\, du = -(1/4)\cos u + C = -(1/4)\cos 4x + C$.

4. With $u = x^3$, $\displaystyle\int x^2 \cot x^3 \csc x^3 dx = (1/3) \int \cot u \csc u\, du = -(1/3)\csc u + C$
    $= -(1/3)\csc x^3 + C$.

7. $\int \dfrac{1}{\cos 2x}\,dx = \int \sec 2x\,dx = \dfrac{1}{2}\ln|\sec 2x + \tan 2x| + C$, (u = 2x could be
    used as in #1).

10. $\int (x + \csc 8x)\,dx = x^2/2 + (1/8)\ln|\csc 8x - \cot 8x| + C$.

13. With u = tan x, du = $\sec^2 x\,dx$, $\displaystyle\int_0^{\pi/4} \tan x \sec^2 x\,dx = \int_0^1 u\,du = 1/2$.

16. $\displaystyle\int_{\pi/6}^{\pi/2} \dfrac{\cos^2 x}{\sin x}\,dx = \int_{\pi/6}^{\pi/2} \dfrac{1 - \sin^2 x}{\sin x}\,dx = \int_{\pi/6}^{\pi/2} (\csc x - \sin x)\,dx = \ln|\csc x - \cot x|$

    $+ \cos x\,]_{\pi/6}^{\pi/2} = 0 - (\ln|2 - \sqrt{3}| + \sqrt{3}/2)$.

19. With u = x + cos x, du = (1 - sin x)dx and $\displaystyle\int \dfrac{1 - \sin x}{x + \cos x}\,dx = \int \dfrac{du}{u} = \ln|u| + C$

    $= \ln|x + \cos x| + C$.

22. $\displaystyle\int_0^{\pi/4} (1 + \sec x)^2\,dx = \int_0^{\pi/4} (1 + 2\sec x + \sec^2 x)\,dx = [x + 2\ln|\sec x + \tan x|$

    $+ \tan x\,]_0^{\pi/4} = \pi/4 + 2\ln|\sqrt{2}+1| + 1$.

25. Since $\sin x = \dfrac{1}{\csc x}$, $\displaystyle\int \dfrac{e^{\cos x}}{\csc x}\,dx = \int e^{\cos x}\sin x\,dx$. Let u = cos x, du =

    -sin x and the integral is $-\displaystyle\int e^u\,du = -e^u + C = -e^{\cos x} + C$.

28. Let u = 1 + 3 cos x, du = -3 sin x. Then 2 sin x dx = -(2/3)du, and

    $\displaystyle\int \dfrac{2\sin x\,dx}{1 + 3\cos x} = -\dfrac{2}{3}\int \dfrac{1}{u}\,du = -\dfrac{2}{3}\ln|u| + C = -\dfrac{2}{3}\ln|1 + 3\cos x| + C$.

31. For x ε [-π/4,π/4], x $\le$ π/4 < 1 $\le$ sec x. Thus A = $\displaystyle\int_{-\pi/4}^{\pi/4} (\sec x - x)\,dx =$

    $\ln|\sec x + \tan x| - \dfrac{x^2}{2}\,]_{-\pi/4}^{\pi/4} = \ln|\sqrt{2} + 1| - \ln|\sqrt{2} - 1|$.

34. $V = \pi\displaystyle\int_0^{\pi} \sin^2 x\,dx = \dfrac{\pi}{2}\int_0^{\pi} (1 - \cos 2x)\,dx = \dfrac{\pi}{2}[x - \dfrac{1}{2}\sin 2x]_0^{\pi} = \dfrac{\pi^2}{2}$.

37. 1st way: let u = tan x, du = $\sec^2 x\,dx$.
    2nd way: let u = sec x, du = sec x tan x dx.
    Since $\sec^2 x = \tan^2 x + 1$, the two answers differ only by a constant.

40. (a) $f'(x) = \sec^2 x \implies L = \displaystyle\int_0^{\pi/4} \sqrt{1 + \sec^4 x}\,dx$.

    (b) With a = 0, b = π/4, n = 4, the partitioning points are 0, π/16, 2π/16 =

$\pi/8$, $3\pi/16$ and $\pi/4$. If $g(x) = \sqrt{1 + \sec^4 x}$, $L \approx \dfrac{(\pi/4-0)}{3(4)}$ $[g(0) + 4g(\pi/16) +$

$2g(\pi/8) + 4g(3\pi/16) + g(\pi/4)]$.

(c) $L \approx (\pi/48)[1.414 + 4(1.442) + 2(1.540) + 4(1.758) + 2.236] \approx$
$(.06545)(19.530) \approx 1.278$.

EXERCISES 9.4, page 430

1. $\sin^{-1}(\pm\sqrt{3}/2) = \pm\pi/3$ since $\sin(\pm\pi/3) = \pm\sqrt{3}/2$ and both angles are in $[-\pi/2,\pi/2]$.

4. (a) $\arcsin(-1) = -\pi/2$ since $\sin(-\pi/2) = -1$ and $-\pi/2 \in [-\pi/2,\pi/2]$.
    (b) $\cos^{-1}(-1) = \pi$ since $\cos \pi = -1$ and $\pi \in [0,\pi]$.

7. $\sin(\cos^{-1}(\sqrt{3}/2)) = \sin \pi/6 = 1/2$.

10. $\tan(\tan^{-1}10) = 10$ since $\tan x$, $\tan^{-1}x$ are inverse functions.

13. Let $a = \sin^{-1}3/5$. Then $\sin a = 3/5$, $\cos a = \sqrt{1-9/25} = 4/5$. Let $b = \tan^{-1}4/3$.
    Then $\sin b = 4/5$, $\cos b = 3/5$. $\cos(\sin^{-1}3/5 + \tan^{-1}4/3) = \cos(a+b) =$
    $\cos a \cos b - \sin a \sin b = (\frac{4}{5})(\frac{3}{5}) - (\frac{3}{5})(\frac{4}{5}) = 0.$

16. Let $a = \sin^{-1}8/17$. Then $\sin a = 8/17$, $\cos a = \sqrt{1-(8/17)^2} = \dfrac{15}{17}$. Then
    $\cos(2 \sin^{-1}8/17) = \cos(2a) = \cos^2 a - \sin^2 a = (\frac{15}{17})^2 - (\frac{8}{17})^2 = \dfrac{161}{289}$.

19. The method of Examples 2 and 3 can be used: let $y = \tan^{-1}x$ so that $x =$
    $\tan y = \sin y/\sqrt{1 - \sin^2 y}$, and now solve for $\sin y$. An alternate method
    is to sketch a right triangle with y as interior
    angle whose tangent is x. We can do this by mak-
    ing the opposite side of length x and the adjacent
    side of length 1. Then the hypotenuse is $\sqrt{x^2+1}$
    and $\sin(\tan^{-1}x) = \sin y = $ opposite/hypotenuse
    $= x/\sqrt{x^2+1}$.

    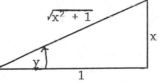

22. Let $y = \tan^{-1}x$. Then $\cos 2y = \cos^2 y - \sin^2 y = (1/\sqrt{x^2+1})^2 - (x/\sqrt{x^2+1})^2 =$
    $(1-x^2)/(x^2+1)$. (Refer to the sketch for #19.)

25. Let $y = \arctan x$. Then the right side of the desired identity is $2y$. Now,
    from #19 above, $\sin y = \sin(\arctan x) = x/\sqrt{x^2+1}$. By the figure there, $\cos y$
    $= 1/\sqrt{x^2+1}$. Thus $2 \sin y \cos y = 2x/(x^2+1)$, and the left side of the identity
    is $\arcsin(2x/(x^2+1)) = \arcsin(2 \sin y \cos y) = \arcsin(\sin 2y) = 2y$, the same
    as the right side.

28. $\cos(\arccos(-x)) = -x$ and, from the formula $\cos(\pi-\theta) = -\cos \theta$ we obtain
    $\cos(\pi-\arccos x) = -\cos(\arccos x) = -x$. The identity is thus established
    since $\cos x$ is one-to-one on $[0,\pi]$. (See Definition (9.29).)

34.

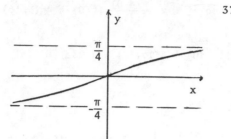

37. (Graph)

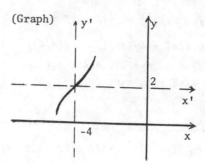

37. Introducing a new x'y' coordinate system with origin at the point (-4,2), we
     have from (7.20), x' = x + 4, y' = y - 2, and the original equation becomes
     y' = $\sin^{-1}$x' which yields the graph shown above.

40. If x = 0, the left side is $0^2 + (\pi/2)^2$, which is definitely not equal to 1.

EXERCISES 9.5, page 436

1. $f'(x) = \dfrac{D_x(3x-5)}{1 + (3x-5)^2} = \dfrac{3}{9x^2-30x+26}$.

2. $f'(x) = \dfrac{D_x(x/3)}{\sqrt{1 - (x/3)^2}} = \dfrac{1/3}{\sqrt{1 - x^2/9}} = \dfrac{1}{\sqrt{9 - x^2}}$.

4. $f'(x) = 2x/(1 + (x^2)^2) = 2x/(1+x^4)$.

7. $f'(x) = x^2 \cdot \dfrac{2x}{1+x^4} + 2x \arctan(x^2)$.

10. $f'(x) = x^2 \cdot \dfrac{5}{5x\sqrt{(5x)^2-1}} + 2x \sec^{-1}5x = x/\sqrt{25x^2-1} + 2x \sec^{-1}5x$.

13. $f'(x) = D_x(\sin^{-1}x)^{-1} = -(\sin^{-1}x)^{-2} D_x\sin^{-1}x = -1/\sqrt{1-x^2} (\sin^{-1}x)^2$.

16. $f'(x) = 4(\dfrac{1}{x} - \arcsin\dfrac{1}{x})^3 D_x(\dfrac{1}{x} - \arcsin\dfrac{1}{x})$. The last derivative is

   $- \dfrac{1}{x^2} - \dfrac{D_x(1/x)}{\sqrt{1-1/x^2}} = - \dfrac{1}{x^2}(1 - \dfrac{1}{\sqrt{1-1/x^2}})$.

19. $f'(x) = \sqrt{x} D_x \sec^{-1}\sqrt{x} + \sec^{-1}\sqrt{x} D_x \sqrt{x} = \dfrac{\sqrt{x} D_x \sqrt{x}}{\sqrt{x} \sqrt{(\sqrt{x})^2-1}} + \dfrac{\sec^{-1}\sqrt{x}}{2\sqrt{x}}$

   $= \dfrac{1/2\sqrt{x}}{\sqrt{x - 1}} + \dfrac{\sec^{-1}\sqrt{x}}{2\sqrt{x}}$.

22. $f'(x) = x D_x \arccos\sqrt{4x+1} + \arccos\sqrt{4x+1} D_x x = \dfrac{-x D_x\sqrt{4x+1}}{\sqrt{1 - \sqrt{4x+1}^2}} + \arccos\sqrt{4x+1}$

   $= \dfrac{-x(4/2\sqrt{4x+1})}{\sqrt{-4x}} + \arccos\sqrt{4x+1} = -2x/\sqrt{4x+1} \sqrt{-4x} + \arccos\sqrt{4x+1}$.

   (Note that the domain of f is [-1/4,0], so there's no problem with $\sqrt{-4x}$.)

25. $x^2 + x \sin^{-1}y = ye^x \implies 2x + xy'/\sqrt{1-y^2} + \sin^{-1}y = ye^x + y'e^x \implies$

$(x/\sqrt{1-y^2} - e^x)y' = ye^x - 2x - \sin^{-1}y$ which yields the given answer.

28. With $u = e^x$, $du = e^x dx$, $\displaystyle\int_0^1 \frac{e^x}{1+e^{2x}} dx = \int_1^e \frac{du}{1+u^2} = \tan^{-1}u \,]_1^e = \tan^{-1}e - \tan^{-1}1$

$= \tan^{-1}e - \pi/4$.

31. With $u = \cos x$, $du = -\sin x\, dx$, $\displaystyle\int \frac{\sin x}{\cos^2 x + 1} dx = -\int \frac{du}{u^2+1} = -\tan^{-1}u + C$

$= -\tan^{-1}(\cos x) + C$.

34. First, $\displaystyle\int \frac{dx}{e^x\sqrt{1-e^{-2x}}} = \int \frac{e^{-x}}{\sqrt{1 - e^{-2x}}} dx$. Let $u = e^{-x}$ so that this becomes

$-\displaystyle\int \frac{du}{\sqrt{1-u^2}} = -\sin^{-1}u + C = -\sin^{-1}(e^{-x}) + C$.

37. First multiply and divide by x^2 so that $\displaystyle\int \frac{1}{x\sqrt{x^6-4}} dx = \int \frac{x^2}{x^3\sqrt{x^6-4}} dx$.

Let $u = x^3$, $du = 3x^2\, dx$, $x^2 dx = (1/3)du$. The integral becomes $\dfrac{1}{3}\displaystyle\int \frac{1}{u\sqrt{u^2-4}} du$

$= \dfrac{1}{3} \cdot \dfrac{1}{2} \sec^{-1}\dfrac{u}{2} + C = \dfrac{1}{6} \sec^{-1}\dfrac{x^3}{2} + C$.

40. We use a slightly different substitution, or change of variable, technique.
Let $u = \sqrt{x}$. Then $x = u^2$ and $dx = 2u\, du$, and

$\displaystyle\int \frac{1}{x\sqrt{x-1}} dx = \int \frac{2u}{u^2\sqrt{u^2-1}} du = 2\int \frac{1}{u\sqrt{u^2-1}} du = 2 \sec^{-1}u + C = 2 \sec^{-1}\sqrt{x} + C$.

43. $A = \displaystyle\int_{-2}^{2} \frac{4}{\sqrt{16-x^2}} dx = 4 \sin^{-1}\frac{x}{4} \,]_{-2}^{2} = 4[\frac{\pi}{6} - (-\frac{\pi}{6})] = \frac{4\pi}{3}$.

46. With $f(x) = \arcsin x$, $x = .25$, $\Delta x = .01$, $\Delta f \approx df = f'(.25)(.01) =$

$\dfrac{1}{\sqrt{1-.25^2}} (.01) = \dfrac{.01}{\sqrt{1 - 1/16}} = .04/\sqrt{15}$.

49. If A and B are the angles shown, between the
horizontal and the top and bottom, respectively,
of the billboard, then $\theta = A-B$ is to be
maximized. If the viewer is x feet from the
base of the building then $\theta = \tan^{-1}(80/x)$

$- \tan^{-1}(60/x)$ and $\dfrac{d\theta}{dx} = \dfrac{-80/x^2}{1+(80/x)^2} - \dfrac{-60/x^2}{1+(60/x)^2}$

$= \dfrac{60}{x^2 + 3600} - \dfrac{80}{x^2 + 6400}$; $\dfrac{d\theta}{dx} = 0$ if $60(x^2 + 6400)$

$- 80(x^2 + 3600) = 0$ or $96,000 - 20x^2 = 0$. Thus

$x = \sqrt{4800} = 40\sqrt{3} \approx 69.3$ feet.

52. Differentiate the right side using the chain rule.

55. $y' = 1/(1+x^2)$ and $y'' = -2x/(1+x^2) > 0$ if $x < 0$ and < 0 if $x > 0$. Thus CU on
     $(-\infty,0)$ and CD on $(0,\infty)$ with a PI at $x = 0$.

58. Recall that if s is arc length of the graph of $y = f(x)$ then $\Delta s \approx ds =$
     $\sqrt{1 + f'(x)^2}\ \Delta x$. Here $f'(x) = 1/(1+x^2)$, $x = 0$ and $\Delta x = 0.1$. $f'(0) = 1$ so
     that $\Delta s \approx \sqrt{1 + f'(0)^2}\ \Delta x = \sqrt{2}(0.1) \approx 0.14$.

EXERCISES 9.6, page 444

1. and 4. These are immediate from the definitions and some easy calculations.

7. $2 \sinh x \cosh x = (2/4)(e^x - e^{-x})(e^x + e^{-x}) = (1/2)(e^{2x} - e^{-2x}) = \sinh 2x$.

10. $\tanh 2x = \dfrac{\sinh 2x}{\cosh 2x} = \dfrac{2 \sinh x \cosh x}{\cosh^2 x + \sinh^2 x}$ (by #7,8)

    $= \dfrac{2 \tanh x}{1 + \tanh^2 x}$ (dividing numerator and denominator by $\cosh^2 x$).

13. $2 \sinh \dfrac{x+y}{2} \cosh \dfrac{x-y}{2} = 2\ \dfrac{(e^{\frac{x+y}{2}} - e^{\frac{-x+y}{2}})}{2}\ \dfrac{(e^{\frac{x-y}{2}} + e^{\frac{-x-y}{2}})}{2}$. When the first terms

    in each parentheses are multiplied together, we get: $e^{\frac{x+y}{2}} e^{\frac{x-y}{2}} = e^{\frac{x+y+x-y}{2}} =$

    $e^{\frac{2x}{2}} = e^x$. Similarly, the other three products can be simplified, and we ob-

    tain for the entire expression: $(e^x + e^y - e^{-y} - e^{-x})/2 =$

    $(e^x - e^{-x})/2 + (e^y - e^{-y})/2 = \sinh x + \sinh y$, as desired.

16. $f'(x) = \sinh\sqrt{4x^2+3}\ D_x\sqrt{4x^2+3} = \dfrac{(\sinh\sqrt{4x^2+3})(8x)}{2\sqrt{4x^2+3}}$.

19. $f'(x) = \dfrac{(x^2+1)(-\text{sech } x^2 \tanh x^2)D_x x^2 - (\text{sech } x^2)D_x(x^2+1)}{(x^2+1)^2}$

    $= \dfrac{-[(x^2+1)(\text{sech } x^2 \tanh x^2)2x + (\text{sech } x^2)(2x)]}{(x^2+1)^2}$.

22. $f'(x) = \cosh(x^2+1)\ D_x(x^2+1) = 2x \cosh(x^2+1)$.

25. $f'(x) = \dfrac{D_x \sinh 2x}{\sinh 2x} = \dfrac{2 \cosh 2x}{\sinh 2x} = 2 \coth 2x$.

28. $f'(x) = (1/2)(\text{sech } 5x)^{-1/2}\ D_x \text{ sech } 5x = -(5 \text{ sech } 5x \tanh 5x)/2\sqrt{\text{sech } 5x}$.

31. $\sinh xy = ye^x \implies (\cosh xy)(xy' + y) = ye^x + y'e^x \implies (x \cosh xy - e^x)y' =$
     $y(e^x - \cosh xy)$.

34. With $u = \ln x$, $du = (1/x)dx$, $\displaystyle\int \dfrac{\cosh \ln x}{x}\ dx = \int \cosh u\ du = \sinh u + C$
     $= \sinh \ln x + C$.

37. With $u = \sinh x$, $\int \sinh x \cosh x \, dx = \int u \, du = u^2/2 + C = (\sinh^2 x)/2 + C.$

40. Let $u = \cosh x$, $du = \sinh x \, dx$. Then $\int \sinh x\sqrt{\cosh x} \, dx = \int \sqrt{u} \, du = \frac{2}{3}u^{3/2}$

$+ C = \frac{2}{3}(\cosh x)^{3/2} + C.$

43. Let $u = 1 - 2 \tanh x$, $du = -2 \, \text{sech}^2 x \, dx$. Then $\int \dfrac{\text{sech}^2 x}{1 - 2 \tanh x} \, dx = -\dfrac{1}{2}\int \dfrac{du}{u} =$

$-\dfrac{1}{2} \ln|u| + C = -\dfrac{1}{2} \ln|1 - 2 \tanh x| + C.$

46. $L_0^1 = \displaystyle\int_0^1 \sqrt{1 + (D_x \cosh x)^2} \, dx$ 52.

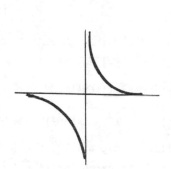

$= \displaystyle\int_0^1 \sqrt{1 + \sinh^2 x} \, dx = \int_0^1 \cosh x \, dx$

$= \sinh x \, \Big]_0^1 = \sinh 1 \approx 1.175.$

EXERCISES 9.7, page 448

1. $y = \cosh^{-1} x \Longrightarrow x = \cosh y = (e^y + e^{-y})/2$, $(x \geq 1)$, $\Longrightarrow e^{2y} - 2xe^y + 1 = 0$

$\Longrightarrow e^y = (2x \underline{+\sqrt{4x^2-4}})/2 \Longrightarrow e^y = x + \sqrt{x^2-1} \longrightarrow$ (9.54). (The + sign must be

chosen to guarantee that $x \to \infty \Longleftrightarrow y \to \infty$.)

4. $D_x(\tanh^{-1} u) = D_u(\frac{1}{2} \ln \frac{1+u}{1-u}) \, D_x u$ 10.

$= \frac{1}{2} D_u(\ln(1+u) - \ln(1-u)D_x u$

$= \frac{1}{2}(\frac{1}{1+u} + \frac{1}{1-u})D_x u = \frac{1}{1-u^2} D_x u.$

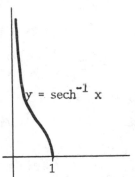

$y = \text{sech}^{-1} x$

13. $f'(x) = \dfrac{D_x \sqrt{x}}{\sqrt{(\sqrt{x})^2 - 1}} = 1/2 \sqrt{x} \sqrt{x-1}.$

16. $f'(x) = \dfrac{D_x \sin 3x}{1 - \sin^2 3x} = \dfrac{3 \cos 3x}{\cos^2 3x} = 3 \sec 3x.$

19. $f'(x) = \dfrac{D_x \cosh^{-1} 4x}{\cosh^{-1} 4x} = 4/(\sqrt{16x^2-1} \, \cosh^{-1} 4x).$

22. With $u = 4x$, $\int \dfrac{1}{\sqrt{16x^2-9}} \, dx = \dfrac{1}{4}\int \dfrac{du}{\sqrt{u^2-9}} = (1/4)\cosh^{-1}(\frac{u}{3}) + C =$

$(1/4)\cosh^{-1}(4x/3) + C.$

25. With $u = e^x$, $\displaystyle\int \frac{e^x}{\sqrt{e^{2x}-16}}\,dx = \int \frac{du}{\sqrt{u^2-16}} = \cosh^{-1}(u/4) + C = \cosh^{-1}(e^x/4) + C.$

28. First multiply and divide the integrand by e^x. Then let $u = e^x$, $du = e^x dx$

and $\displaystyle\int \frac{1}{\sqrt{5-e^{2x}}}\,dx = \int \frac{e^x dx}{e^x\sqrt{5-e^{2x}}} = \int \frac{du}{u\sqrt{5-u^2}} = -\frac{1}{\sqrt{5}}\,\text{sech}^{-1}\,\frac{|u|}{5} + C = -\frac{1}{\sqrt{5}}\,\text{sech}^{-1}\,\frac{e^x}{5}$

+ C.   ($|e^x| = e^x$ since $e^x > 0$.)

EXERCISES 9.8 (Review), page 449

1. $f'(x) = -\sin\sqrt{3x^2+x}\ D_x\sqrt{3x^2+x} = (-\sin\sqrt{3x^2+x})(\tfrac{1}{2})(3x^2+x)^{-1/2}\ D_x(3x^2+x)$

    $= (-\sin\sqrt{3x^2+x})(6x+1)/2\sqrt{3x^2+x}.$

4. $f'(x) = D_x(x^3 + \csc 6x)^{1/3} = (1/3)(x^3 + \csc 6x)^{-2/3}\ D_x(x^3 + \csc 6x) =$

    $(1/3)(x^3 + \csc 6x)^{-2/3}(3x^2 - 6\csc 6x \cot 6x)$ or

    $(3x^2 - 6\csc 6x \cot 6x)/3\sqrt[3]{(x^3 + \csc 6x)^2}.$

7. $f'(x) = (\sin^{-1} 5x\ D_x(3x+7)^4 - (3x+7)^4\ D_x \sin^{-1} 5x)/(\sin^{-1} 5x)^2$

    $= [(\sin^{-1} 5x)4(3x+7)^3 3 - (3x+7)^4 \cdot \dfrac{5}{\sqrt{1-25x^2}}]/(\sin^{-1} 5x)^2.$

10. $f'(x) = 7^{\sin 3x} \ln 7\ D_x \sin 3x = 7^{\sin 3x} \ln 7(3\cos 3x).$

13. $f(x) = \ln(\csc^3 2x) = 3\ln \csc 2x \Longrightarrow f'(x) = 3 \cdot \dfrac{D_x \csc 2x}{\csc 2x} =$

    $3 \cdot \dfrac{-(\csc 2x \cot 2x)(2)}{\csc 2x} = -6\cot 2x.$

16. $f(x) = \cot(1/x) + 1/\cot x = \cot(1/x) + \tan x \Longrightarrow f'(x) = -\csc^2(1/x)D_x(1/x)$

    $+ \sec^2 x = (1/x^2)\csc^2(1/x) + \sec^2 x.$

19. $f'(x) = \sinh e^{-5x}\ D_x e^{-5x} = (\sinh e^{-5x})(-5e^{-5x}).$

22. $f'(x) = \sec(\sec x)\tan(\sec x)D_x \sec x = \sec(\sec x)\tan(\sec x)(\sec x \tan x).$

25. $f'(x) = 3\sin^2 e^{-2x}\ D_x \sin e^{-2x} = 3\sin^2 e^{-2x} \cos e^{-2x} D_x(e^{-2x}) =$

    $3\sin^2 e^{-2x} \cos e^{-2x}(-2e^{-2x}).$

28. $f'(x) = \sec 5x\ D_x \tan 5x + \tan 5x\ D_x \sec 5x = 5\sec^3 5x + 5\tan^2 5x \sec 5x.$

31. $f'(x) = \dfrac{D_x\sqrt{1-x^2}}{\sqrt{1-(\sqrt{1-x^2})^2}} = \dfrac{-x}{\sqrt{1-x^2}\ \sqrt{x^2}} = \pm\dfrac{1}{\sqrt{1-x^2}}$, + if $x < 0$, - if $x > 0$.

34. $f'(x) = \frac{1}{2}(\sin \sqrt{x})^{-1/2} D_x \sin \sqrt{x} = \frac{\cos \sqrt{x} \, D_x \sqrt{x}}{2\sqrt{\sin \sqrt{x}}} = \frac{\cos \sqrt{x}}{4\sqrt{x}\sqrt{\sin \sqrt{x}}}$.

37. $f'(x) = \frac{D_x \tan^{-1}x}{1+(\tan^{-1}x)^2} = 1/(1+x^2)[1 + (\tan^{-1}x)^2]$.

40. $f'(x) = \frac{D_x \tanh(5x+1)}{\tanh(5x+1)} = \frac{5 \ \text{sech}^2(5x+1)}{\tanh \ (5x+1)}$.

43. $f'(x) = (D_x x^2)/\sqrt{(x^2)^2+1} = 2x/\sqrt{x^4+1}$.

46. With $u = x/2$, $2 \ du = dx$, $\int \csc(x/2)\cot(x/2)dx = 2\int \csc u \cot u \ du =$
    $-2 \csc u + C = -2 \csc(x/2) + C$.

49. With $u = 9x$, $dx = (1/9)du$, $\int (\cot 9x + \csc 9x)dx = (1/9)\int(\cot u + \csc u)du =$
    $(1/9)(\ln|\sin u| + \ln|\csc u - \cot u|) + C$, where $u = 9x$. (9.23) and (9.25) were
    used for the integrals.

52. $\int \cot 2x \csc 2x \ dx = -(1/2)\csc 2x + C$.

55. $\int \frac{\sin 4x}{\tan 4x} dx = \int \cos 4x \ dx = (1/4)\sin 4x + C$.

58. $\int \frac{1}{4 + 9x^2} dx = \frac{1}{9} \int \frac{dx}{(4/9) + x^2} = \frac{1}{9} \cdot \frac{3}{2} \tan^{-1} \frac{x}{2/3} + C = \frac{1}{6}\tan^{-1}(\frac{3x}{2}) + C$.

61. With $u = x^2$, $\int \frac{x}{\text{sech } x^2} dx = \frac{1}{2} \int \cosh u \ du = \frac{1}{2} \sinh u + C = \frac{1}{2} \sinh(x^2) + C$.

64. $\int_0^{\pi/2} \frac{\cos x}{1 + \sin^2 x} dx = \tan^{-1}(\sin x)]_0^{\pi/2} = \tan^{-1}1 - \tan^{-1}0 = \frac{\pi}{4}$.

67. $u = 2 + \cot x \Rightarrow du = -\csc^2 x \ dx$. $\int \frac{\csc^2 x}{2 + \cot x} dx = -\int \frac{du}{u} = -\ln|u| + C =$
    $-\ln|2 + \cot x| + C$.

70. $u = 1 - 2x \Rightarrow dx = -(1/2)du$, $\int \text{sech}^2(1-2x)dx = -(1/2)\int \text{sech}^2 u \ du =$
    $-(1/2)\tanh u + C = -(1/2)\tanh(1-2x) + C$.

73. $u = 2x \Rightarrow dx = (1/2)du$ and $x = u/2$. $\int \frac{1}{x\sqrt{9-4x^2}} dx = \int \frac{(1/2)du}{(u/2)\sqrt{9-u^2}} =$

    $\int \frac{1}{u\sqrt{3^2-u^2}} du = -\frac{1}{3} \text{sech}^{-1} \frac{|u|}{3} + C = -\frac{1}{3} \text{sech}^{-1} \frac{|2x|}{3} + C$.

76. Instead of changing variables, we proceed as follows. $\int \frac{1}{\sqrt{25x^2+36}} dx =$

    $\int \frac{1}{\sqrt{25(x^2+36/25)}} dx = \frac{1}{5} \int \frac{1}{\sqrt{x^2+(6/5)^2}} dx = \frac{1}{5} \sinh^{-1} \frac{x}{6/5} + C = \frac{1}{5} \sinh^{-1} \frac{5x}{6} + C$.

79. m_{AB} = (7+3)/(4-2) = 5. y' = $3/\sqrt{1-9x^2}$ = 5 if 9 = 25(1-9x²) or x² = 16/225.

   Thus x = ±4/15 and the points are (±4/15,sin⁻¹(±4/5)).

82. a(t) = s"(t) = -k²(a sin kt + b cos kt). The distance from the origin is

   |s(t)|, and we see that |a(t)| = k²|s(t)|.

85. y' = $x/\sqrt{1-x^2}$ + sin⁻¹x and y" = $\dfrac{\sqrt{1-x^2} + x^2/\sqrt{1-x^2}}{1 - x^2}$ + $\dfrac{1}{\sqrt{1-x^2}}$ = $\dfrac{2 - x^2}{(1-x^2)^{3/2}}$ > 0

   on (-1,1). Thus the graph is CU on (-1,1).

88. Let θ be the bottom angle, w the width at the top, and
   h the depth. We seek θ which maximizes the area. The
   volume of the trough will then be maximized since it is
   proportional to the cross-sectional area. By elementary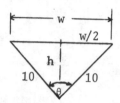
   trigonometry $\dfrac{w}{2}$ = 10 sin $\dfrac{\theta}{2}$, h = 10 cos $\dfrac{\theta}{2}$. Thus A =

   (1/2)wh = 100 sin $\dfrac{\theta}{2}$ cos $\dfrac{\theta}{2}$ = 50 sin θ, obviously maximized

   when sin θ = 1, or θ = π/2. No differentiation required!

EXERCISES 10.1, page 458

1. $u = x$, $dv = e^{-x}dx \implies du = dx$, $v = -e^{-x}$. $\int xe^{-x}dx = -xe^{-x} - \int (-e^{-x})dx =$
    $-(x+1)e^{-x} + C$.

4. $u = x^2$, $dv = \sin 4x\, dx \implies du = 2x\, dx$, $v = -(1/4)\cos 4x$, $I = \int x^2 \sin 4x\, dx$
    $= x^2(-1/4)\cos 4x - \int (-1/4)\cos 4x(2x)dx = -(x^2/4)\cos 4x + (1/2)\int x \cos 4x\, dx$.
    Another integration by parts is necessary. $u = x$, $dv = \cos 4x\, dx \implies du = dx$,
    $dv = (1/4)\sin 4x$, and $\int x \cos 4x\, dx = (x/4)\sin 4x - (1/4)\int \sin 4x\, dx =$
    $(x/4)\sin 4x + (1/16)\cos 4x$. Combining results we obtain:
    $I = -(x^2/4)\cos 4x + (1/2)[(x/4)\sin 4x + (1/16)\cos 4x] + C$.

7. $u = x$, $dv = \sec x \tan x\, dx \implies du = dx$, $v = \sec x$. $\int x \sec x \tan x\, dx =$
    $x \sec x - \int \sec x\, dx = x \sec x - \ln|\sec x + \tan x| + C$.

10. $u = x^3$, $dv = e^{-x}dx \implies du = 3x^2 dx$, $v = -e^{-x}$. $I = \int x^3 e^{-x}dx = -x^3 e^{-x} +$
    $3\int x^2 e^{-x}dx$. Now, $u = x^2$, $dv = e^{-x}dx \implies du = 2x\, dx$, $v = -e^{-x}$. So, $I =$
    $-x^3 e^{-x} + 3[-x^2 e^{-x} + 2\int xe^{-x}dx]$. The last integral was done in #1. So, combining all, $I = -(x^3 + 3x^2 + 6x + 6)e^{-x} + C$.

13. $u = \ln x$, $dv = \sqrt{x}\, dx \implies du = \frac{1}{x}dx$, $v = \frac{2}{3}x^{3/2}$. $\int \sqrt{x} \ln x\, dx = \frac{2}{3}x^{3/2} \ln x -$
    $\frac{2}{3}\int \frac{x^{3/2}}{x}\, dx = \frac{2}{3}x^{3/2} \ln x - \frac{4}{9}x^{3/2} + C$. N.B. $\int \frac{x^{3/2}}{x}dx = \int x^{1/2}dx = \frac{2}{3}x^{3/2} + D$.

16. $u = \tan^{-1}x$, $dv = x\, dx \implies du = \frac{1}{1+x^2}\, dx$, $v = \frac{x^2}{2}$. $\int x \tan^{-1}x\, dx = \frac{x^2}{2}\tan^{-1}x -$
    $\frac{1}{2}\int \frac{x^2}{1+x}\, dx = \frac{x^2}{2}\tan^{-1}x - \frac{1}{2}\int (1 - \frac{1}{1+x^2})dx = \frac{x^2}{2}\tan^{-1}x - \frac{x}{2} + \frac{1}{2}\tan^{-1}x + C$.
    N.B. $\frac{x^2}{1+x^2} = 1 - \frac{1}{1+x^2}$ by ordinary long division.

19. With $y = \cos x$, $dy = -\sin x\, dx$, $\int \sin x \ln \cos x\, dx = -\int \ln y\, dy = -(y \ln y - y)$
    $+ C$ (by Example 3) $= -\cos x \ln \cos x + \cos x + C$.

21. $u = \sec x$, $dv = \sec^2 x\, dx \implies du = \sec x \tan x\, dx$, $v = \tan x$. Then I =
    $\int \sec^3 x\, dx = \sec x \tan x - \int \sec x \tan^2 x\, dx = \sec x \tan x - \int \sec^3 x\, dx +$
    $\int \sec x\, dx$. (Recall: $\tan^2 x = \sec^2 x - 1$.) Thus $2I = \sec x \tan x +$
    $\ln|\sec x + \tan x| + C$.

22. $u = \csc^3 x$, $dv = \csc^2 x\, dx \implies du = -3 \csc^3 x \cot x\, dx$, $v = -\cot x$. After

using $\cot^2 x = \csc^2 x - 1$, this yields $4\int \csc^5 x \, dx = -\cot x \csc^3 x + 3 \cdot$

$\int \csc^3 x \, dx.$ The last integral is evaluated using the procedure of #21 but with

$u = \csc x$, $dv = \csc^2 x \, dx$. Ultimately, $\int \csc^5 x \, dx = \frac{1}{8}[3 \ln|\csc x - \cot x| -$

$\cot x \csc x (3 + 2\csc^2 x)] + C.$

25. $u = x$, $dv = \sin 2x \, dx \implies du = dx$, $v = -\frac{1}{2}\cos 2x$ and $\int_0^{\pi/2} x \sin 2x \, dx =$

$-\frac{x}{2}\cos 2x \Big]_0^{\pi/2} + \frac{1}{2}\int_0^{\pi/2} \cos 2x \, dx = (\frac{\pi}{4} - 0) + \frac{1}{4}\sin 2x \Big]_0^{\pi/2} = \frac{\pi}{4} + 0 = \frac{\pi}{4}.$

28. This is a tricky one to get started correctly. We have to break up the

integrand as follows: $u = x^3$, $dv = \frac{x^2}{\sqrt{1-x^3}}\, dx \implies du = 3x^2 dx$, $v = \int \frac{x^2}{\sqrt{1-x^3}}\, dx.$

To evaluate v, let $y = 1-x^3$, $dy = -3x^2 dx$, $x^2 dx = -(1/3)dy$. Then $v =$

$-\frac{1}{3}\int y^{-1/2}\, du = -(2/3)y^{1/2} = -(2/3)\sqrt{1-x^3}.$ Using these results, $I = \int \frac{x^5}{\sqrt{1-x^3}}dx$

$= -\frac{2}{3}x^3\sqrt{1-x^3} - \frac{2}{3}\int \sqrt{1-x^3}\, 3x^2 dx.$ This last integral can be done with the same

substitution, $y = 1-x^3$, as above to obtain $I = -\frac{2}{3}x^3\sqrt{1-x^3} - \frac{2}{3}\cdot\frac{2}{3}(1-x^3)^{3/2} +$

C. By factoring out $\sqrt{1-x^3}$, and simplifying, this reduces to $-(2/9)\sqrt{1-x^3}(x^3+2)$

+ C.

31. $u = (\ln x)^2$, $dv = dx \implies du = (2 \ln x)\frac{1}{x}\, dx$, $v = x \cdot \int(\ln x)^2 dx = x(\ln x)^2 -$

$2\int(\ln x)\frac{1}{x}\, x \, dx = x(\ln x)^2 - 2\int\ln x \, dx = x(\ln x)^2 - 2(x \ln x - x) + C$ by

the result of Example 3.

34. $u = x+4$, $dv = \cosh 4x \, dx \implies du = dx$, $v - (1/4)\sinh 4x.$ $\int(x+4)\cosh 4x \, dx =$

$(1/4)(x+4)\sinh 4x - (1/4)\int\sinh 4x \, dx = (1/4)(x+4)\sinh 4x - (1/16)\cosh 4x + C.$

37. $u = \cos^{-1} x$, $dv = dx \implies du = -1/\sqrt{1-x^2}\, dx = -(1-x^2)^{-1/2}dx$, $v = x.$ $\int \cos^{-1} x \, dx$

$= x \cos^{-1} x - \int(1-x^2)^{-1/2}(-x)dx = x \cos^{-1} x - (1-x^2)^{1/2} + C.$

40. Let $u = x^m$, $dv = \sin x \, dx$, and the formula follows immediately.

43. $\int x^5 e^x dx = x^5 e^x - 5\int x^4 e^x dx = x^5 e^x - 5[x^4 e^x - 4\int x^3 e^x dx] = x^5 e^x - 5x^4 e^x +$

$20[x^3 e^x - 3\int x^2 e^x dx] = x^5 e^x - 5x^4 e^x + 20x^3 e^x - 60[x^2 e^x - 2\int x \, e^x] =$

$x^5 e^x - 5x^4 e^x + 20x^3 e^x - 60x^2 e^x + 120[x \, e^x - \int e^x].$ The last integral is

just $e^x + C.$

45. $A = \int_0^{\pi^2} \sin\sqrt{x}\, dx.$ First, $y = \sqrt{x}$, $dy = (1/2\sqrt{x})dx \implies dx = 2\sqrt{x}\, dy = 2y \, dy$, and

$A = 2\int_0^{\pi} y \sin y \, dy.$ Now, $u = y$, $dv = \sin y \, dy \implies du = dy$, $v = -\cos y$, and

$$A = 2(-y \cos y]_0^\pi + \int_0^\pi \cos y \, dy) = 2(\pi + \sin y]_0^\pi) = 2\pi.$$

46. By the disc method, $V = \pi \int_0^{\pi/2} x^2 \sin x \, dx$. Using #40, $\int x^2 \sin x \, dx =$

$-x^2 \cos x + 2\int x \cos x \, dx$. In the last integral, $u = x$, $dv = \cos x \, dx \implies du$

$= dx$, $v = \sin x$, and $\int x \cos x \, dx = x \sin x - \int \sin x \, dx = x \sin x + \cos x$.

Combining these results: $V = \pi(-x^2 \cos x + 2x \sin x + 2 \cos x)]_0^{\pi/2}$

$= \pi[(0 + 2(\frac{\pi}{2}) + 0) - (0 + 0 + 2)] = \pi(\pi - 2)$.

49. Because of symmetry, we can compute the force
on the right half and double it to get the
total force. The left boundary of this half
is $x = 3\pi/2$. The right boundary is $y =$
$\sin x$, $3\pi/2 \le x \le y$. We must express this
equation in terms of y as the independent
variable. We obtain for such x, $\sin^{-1} y =$
$x - 2\pi$, or $x = \sin^{-1} y + 2\pi$, since, by the
definition, $-\pi/2 \le \sin^{-1} y \le \pi/2$, and here,

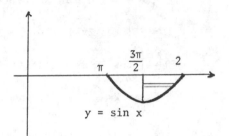

y = sin x

$3\pi/2 \le x \le 2\pi$. Then by (6.16), if F is the force on the right half, $F =$

$$\rho \int_{-1}^0 (0-y)[(\sin^{-1}y + 2\pi) - 3\pi/2] \, dy = -\rho \int_1^0 y(\sin^{-1}y + \pi/2) \, dy = \rho \int_{-1}^0 y \sin^{-1}y \, dy$$

$-\frac{\pi\rho}{2} \int_{-1}^0 y \, dy$. Let $I = \int_{-1}^0 y \sin^{-1}y \, dy$. $u = y$, $dv = \sin^{-1}y \, dy \implies du = dy$,

$v = y \sin^{-1}y + \sqrt{1-y^2}$ (by #12). So, $I = y^2 \sin^{-1}y + y\sqrt{1-y^2}]_{-1}^0 - \int_{-1}^0 (y \sin^{-1}y + \sqrt{1-y^2}) \, dy$

$= (0 - \sin^{-1}(-1)) - I - \int_{-1}^0 \sqrt{1-y^2} \, dy$. We cannot evaluate the last integral

by antidifferentiation yet. However, it is the area of a quarter circle of
radius one. (It's the area of the region in the 4th quadrant enclosed by
$x^2 + y^2 = 1$.) Thus its value is $\pi/4$, and, using $\sin^{-1}(-1) = -\pi/2$ and trans-
posing the I term, we obtain $2I = -(-\pi/2) - \pi/4 = \pi/4$, or $I = \pi/8$. Since the

value of $\int_{-1}^0 y \, dy$ is $-1/2$, we obtain: $F = -\rho(\pi/8) - (\pi\rho/2)(-1/2) =$

$(\pi\rho/4)(1-1/2) = \pi\rho/8$. The total force is thus $\pi\rho/4 = 62.5 \, \pi/4$.

EXERCISES 10.2, page 464

1. $\int \cos^3 x\ dx = \int \cos^2 x \cos x\ dx = \int (1 - \sin^2 x)\cos x\ dx = \int (1-u^2)\ du$ (with

$u = \sin x) = u - \dfrac{u^3}{3} + C = \sin x - \dfrac{\sin^3 x}{3} + C.$

4. $\int \cos^7 x\ dx = \int (1-\sin^2 x)^3 \cos x\ dx = \int (1-u^2)^3 du = \int (1 - 3u^2 + 3u^4 - u^6)\,du$

(with $u = \sin x) = u - u^3 + \dfrac{3}{5}u^5 - \dfrac{1}{7}u^7 + C = \sin x - \sin^3 x + \dfrac{3}{5}\sin^5 x -$

$\dfrac{1}{7}\sin^7 x + C.$

7. $\int \sin^6 x\ dx = \int [\tfrac{1}{2}(1 - \cos 2x)]^3 dx = \dfrac{1}{8}\int (1 - 3\cos 2x + 3\cos^2 2x - \cos^3 2x)\,dx$

$= \dfrac{1}{8}(A - B + C - D)$ where $A = \int 1\ dx = x,\ B = 3\int \cos 2x\ dx = \dfrac{3}{2}\sin 2x,$

$C = 3\int \cos^2 2x\ dx = \dfrac{3}{2}\int (1 + \cos 4x)\,dx = \dfrac{3}{2}(x + (\sin 4x)/4),$

$D = \int \cos^3 2x\ dx = \dfrac{1}{2}\int \cos^3 t\ dt$ (with $t = 2x) = \dfrac{1}{2}(\sin t - \dfrac{\sin^3 t}{3})$ (by Exercise 1)

$= \dfrac{1}{2}(\sin 2x - \dfrac{\sin^3 2x}{3})$. Now combine these to obtain the answer in the text.

10. $\int \sec^6 x\ dx = \int \sec^4 x \sec^2 x\ dx = \int (1 + \tan^2 x)^2 \sec^2 x\ dx = \int (1 + u^2)^2\ du = (u=\tan x)$

$\int (1 + 2u^2 + u^4)\,du = u + \dfrac{2}{3}u^3 + \dfrac{1}{5}u^5 + C = \tan x + \dfrac{2}{3}\tan^3 x + \dfrac{1}{5}\tan^5 x + C.$

11. $\int \tan^3 x \sec^3 x\ dx = \int \tan^2 x \sec^2 x \sec x \tan x\ dx = \int (\sec^2 x - 1)\sec^2 x \sec x \cdot$

$\tan x\ dx = \int (u^2-1)u^2 du = \dfrac{1}{5}u^5 - \dfrac{1}{3}u^3 + C$ $(u = \sec x) = \dfrac{1}{5}\sec^5 x - \dfrac{1}{3}\sec^3 x + C.$

13. $\int \tan^6 x\ dx = \int \dfrac{\tan^6 x}{\sec^2 x}\sec^2 x\ dx = \int \dfrac{\tan^6 x}{\tan^2 x + 1}\sec^2 x\ dx = \int \dfrac{u^6}{u^2+1}\ du$ $(u = \tan x)$

$= \int (u^4 - u^2 + 1 - \dfrac{1}{u^2+1})\,du$ (by long division) $= \dfrac{1}{5}u^5 - \dfrac{1}{3}u^3 + u - \tan^{-1}u + C =$

$\dfrac{1}{5}\tan^5 x - \dfrac{1}{3}\tan^3 x + \tan x - x + C.$

16. $\int \dfrac{\cos^3 x}{\sqrt{\sin x}}\ dx = \int \sin^{-1/2}x(1-\sin^2 x)\ \cos x\ dx = \int (u^{-1/2} - u^{3/2})\,du = 2u^{1/2} -$

$\dfrac{2}{5}u^{5/2} + C$ $(u = \sin x) = 2\sqrt{\sin x} - \dfrac{2}{5}\sqrt{\sin^5 x} + C.$

19. $\displaystyle\int_0^{\pi/4} \sin^3 x\ dx = \int_0^{\pi/4} (1-\cos^2 x)\sin x\ dx = -\cos x + \dfrac{1}{3}\cos^3 x\ \Big]_0^{\pi/4} =$

$(-1 + \dfrac{1}{6})/\sqrt{2} - (- \dfrac{2}{3}) = 2/3 - 5/6\sqrt{2}.$

22. By the product formula, $\cos x \cos 5x = \dfrac{1}{2}(\cos 4x + \cos 6x)$. So,

$$\int_0^{\pi/4} \cos x \cos 5x \, dx = \frac{1}{2} \int_0^{\pi/4} (\cos 4x + \cos 6x) dx = \frac{1}{2}[\frac{\sin 4x}{4} + \frac{\sin 6x}{6}]_0^{\pi/4}$$

$$= \frac{1}{2}[(0 - \frac{1}{6}) - (0-0)] = -1/12.$$

25. $\displaystyle\int \csc^4 x \cot^4 x \, dx = \int \csc^2 x \cot^4 x \csc^2 x \, dx = \int (1 + \cot^2 x) \cot^4 x \csc^2 x \, dx =$

   $\displaystyle -\int (1 + u^2) u^4 du \quad (u = \cot x) \quad = -u^5/5 - u^7/7 + C = -(\cot^5 x)/5 - (\cot^7 x)/7 + C.$

28. $\cos x = 1/\sec x$, $\tan x = \sin x/\cos x \Longrightarrow \dfrac{\tan^2 x - 1}{\sec^2 x} = \sin^2 x - \cos^2 x = -\cos 2x.$

   Thus $\displaystyle\int \frac{\tan^2 - 1}{\sec^2 x} \, dx = -\int \cos 2x \, dx = -\frac{1}{2} \sin 2x + C = -\sin x \cos x + C.$

   (Note that if $\tan x$ and $\sec x$ had been retained, we would have obtained:

   $\displaystyle\int \frac{\tan^2 x - 1}{\sec^2 x} \, dx = \int \frac{\tan^2 x - 1}{\sec^4 x} \sec^2 x \, dx = \int \frac{(\tan^2 x - 1)\sec^2 x}{(\tan^2 x + 1)^2} dx = \int \frac{u^2 - 1}{(u^2 + 1)^2} \, du,$

   an integral which we can not yet evaluate.)

31. $V = \pi \displaystyle\int_0^\pi \sin^2 x \, dx = \frac{\pi}{2} \int_0^\pi (1 - \cos 2x) dx = \frac{\pi}{2}[x - \frac{\sin 2x}{2}]_0^\pi = \frac{\pi}{2}[(\pi - 0) - 0] = \frac{\pi^2}{2}.$

34. $a(t) = \sin^2 t \cos t \Longrightarrow v(t) = \frac{1}{3} \sin^3 t + C.$ Using $v(0) = 10$, we get $C = 10.$

   $v(t) = \frac{1}{3}\sin^3 t + 10 \Longrightarrow s(t) = -\frac{1}{3}\cos t + \frac{1}{9}\cos^3 t + 10t + D$ (using the result

   of Exercise 19). Using $s(0) = 0$, we get $0 = -1/3 + 1/9 + D \Longrightarrow D = 2/9.$

   Thus $s(t) = \frac{1}{9}(-3 \cos t + \cos^3 t + 90t + 2).$

EXERCISES 10.3, page 470

1. Let $x = 2 \sin \theta$, $dx = 2 \cos \theta \, d\theta$, $\sqrt{4-x^2} = 2 \cos \theta.$ $\displaystyle\int \frac{x^2}{\sqrt{4-x^2}} \, dx =$

   $\displaystyle\int \frac{4 \sin^2\theta \ (2 \cos \theta) d\theta}{2 \cos \theta} = 2 \int (1 - \cos 2\theta) d\theta =$

   $2(\theta - \dfrac{\sin 2\theta}{2}) + C = 2(\theta - \sin \theta \cos \theta) + C =$

   $2(\sin^{-1} \frac{x}{2} - \frac{x}{2} \cdot \dfrac{\sqrt{4-x^2}}{2}) + C.$

4. Let $x = 3 \tan \theta$, $dx = 3 \sec^2\theta \, d\theta$, $\sqrt{x^2+9} = 3 \sec \theta.$ $\displaystyle\int \frac{dx}{x^2\sqrt{x^2+9}} =$

   $\displaystyle\int \frac{3 \sec^2\theta \, d\theta}{9 \tan^2\theta \ (3 \sec \theta)} = \frac{1}{9} \int \frac{\sec \theta}{\tan^2\theta} \, d\theta = \frac{1}{9} \int \frac{(1/\cos \theta)}{(\sin \theta/\cos \theta)^2} \, d\theta =$

   $\frac{1}{9} \displaystyle\int \frac{\cos \theta}{\sin^2\theta} \, d\theta = \frac{1}{9} \int \cot \theta \csc \theta \, d\theta =$

   $-\frac{1}{9} \csc \theta + C = -\dfrac{\sqrt{x^2+9}}{9x} + C.$

7. You could use x = 2 sin θ, but the quickest way is the simpler substitution

u = 4-x², du = -2x dx so that $\int \frac{x}{\sqrt{4-x^2}} dx = -\frac{1}{2} \int u^{-1/2} du = -u^{1/2} + C =$

$-\sqrt{4-x^2} + C.$

10. Noting that $\sqrt{4x^2-25} = 2\sqrt{x^2 - 25/4}$, we let $x = \frac{5}{2} \sec θ$, $dx = \frac{5}{2} \sec θ \tan θ\, dθ$,

$\sqrt{x^2 - 25/4} = \frac{5}{2}\tan θ.$ Then $\int \frac{dx}{\sqrt{4x^2-25}} = \frac{1}{2} \int \frac{dx}{\sqrt{x^2 - 25/4}} =$

$\frac{1}{2} \int \frac{\frac{5}{2}\sec θ \tan θ}{\frac{5}{2} \tan θ} dθ = \frac{1}{2} \int \sec θ\, dθ = \frac{1}{2} \ln|\sec θ + \tan θ| + C$

$= \frac{1}{2} \ln\left|\frac{2x}{5} + \frac{\sqrt{4x^2-25}}{5}\right| + C = \frac{1}{2} \ln\left|2x + \sqrt{4x^2-25}\right| + C',$

$C' = C - \frac{1}{2} \ln 5.$

13. Noting that $\sqrt{9-4x^2} = 2\sqrt{9/4 - x^2}$, we let $x = \frac{3}{2} \sin θ$, $dx = \frac{3}{2} \cos θ\, dθ$,

$\sqrt{9/4 - x^2} = \frac{3}{2} \cos θ.$ $\int \sqrt{9-4x^2}\, dx = 2 \int \sqrt{9/4 - x^2}\, dx =$

$2 \int (\frac{3}{2} \cos θ)(\frac{3}{2} \cos θ)dθ = \frac{9}{2} \int \cos^2 θ\, dθ = \frac{9}{4} \int (1 + \cos 2θ)dθ$

$= \frac{9}{4}(θ + \frac{\sin 2θ}{2}) + C = \frac{9}{4}(θ + \sin θ \cos θ) + C =$

$\frac{9}{4}(\sin^{-1}\frac{2x}{3} + (\frac{2x}{3})(\frac{\sqrt{9-4x^2}}{3})) + C.$

16. We could let x = 3 sec θ, but the quickest way is to let u = x²-9, du = 2x dx,

and $\int x\sqrt{x^2-9}\, dx = \frac{1}{2} \int u^{1/2} du = \frac{1}{3}u^{3/2} + C = \frac{1}{3}(x^2-9)^{3/2} + C.$

19. Let $x = \sqrt{3} \sec θ$, $dx = \sqrt{3} \sec θ \tan θ\, dθ$, $\sqrt{x^2-3} = \sqrt{3} \tan θ.$ $\int \frac{1}{x^4\sqrt{x^2-3}} dx$

$= \int \frac{\sqrt{3} \sec θ \tan θ}{9 \sec^4 θ(\sqrt{3} \tan θ)} dθ = \frac{1}{9} \int \frac{1}{\sec^3 θ} dθ = \frac{1}{9}\int \cos^3 θ\, dθ$

$= \frac{1}{9} \int (1 - \sin^2 θ)\cos θ\, dθ = \frac{1}{9} \int (1-u^2)du$ (with u =

$\sin θ$) $= \frac{1}{9}(u - \frac{u^3}{3}) + C = \frac{1}{9}(\sin θ - \frac{\sin^3 θ}{3}) + C$

$= \frac{\sin θ}{27}(3 - \sin^2 θ) + C = \frac{\sqrt{x^2-3}}{27x}(3 - \frac{x^2-3}{x^2}) + C$

$= \frac{\sqrt{x^2-3}(2x^2+3)}{27x^3} + C.$

22. Let $x = \sin θ$, $dx = \cos θ\, dθ$, $\sqrt{1-x^2} = \cos θ.$ $\int \frac{3x-5}{\sqrt{1-x^2}} dx$

$= \int (3 \sin θ - 5)dθ = -3 \cos θ - 5θ + C = -3\sqrt{1-x^2} -$
$5 \sin^{-1}x + C.$

25. This is a special case of #28 with a = 1.

28. Let $u = a \sec \theta$, $du = a \sec \theta \tan \theta \, d\theta$, $\sqrt{u^2 - a^2} = a \tan \theta$. Then $\displaystyle\int \frac{1}{u\sqrt{u^2 - a^2}} \, du$

$\displaystyle = \frac{1}{a} \int d\theta = \frac{1}{a}\theta + C = \frac{1}{a} \sec^{-1} \frac{u}{a} + C.$

31. $\displaystyle L_0^2 = \int_0^2 \sqrt{1 + y'^2} \, dx = \int_0^2 \sqrt{1 + x^2} \, dx.$ Let $x = \tan \theta$, $dx = \sec^2 \theta \, d\theta$, $\sqrt{1 + x^2} = \sec \theta.$

Then $\displaystyle\int \sqrt{1 + x^2} \, dx = \int \sec^3 \theta \, d\theta = \frac{1}{2}(\sec \theta \tan \theta +$

$\ln|\sec \theta + \tan \theta|\,) + C$ by Exercise 21, Sec. 10.1.

Thus $\displaystyle L_0^2 = \frac{1}{2}[x\sqrt{x^2+1} + \ln|\sqrt{x^2+1} + x|]\Big|_0^2 =$

$\displaystyle \frac{1}{2}[2\sqrt{5} + \ln|\sqrt{5} + 2| - 0] \approx 2.46.$

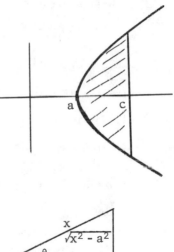

34. The upper and lower boundaries are $y = \pm\frac{b}{a}\sqrt{x^2 - a^2}$

for $a \le x \le c$. $\displaystyle A = 2\frac{b}{a} \int_a^c \sqrt{x^2 - a^2} \, dx.$ Let $x =$

$a \sec \theta$, $dx = a \sec \theta \tan \theta \, d\theta$, $\sqrt{x^2 - a^2} = a \tan \theta$.

Then $\displaystyle\int \sqrt{x^2 - a^2} \, dx = \int a \tan \theta \, (a \sec \theta \tan \theta) \, d\theta =$

$\displaystyle a^2 \int \tan^2 \theta \sec \theta \, d\theta = a^2 \int (\sec^3 \theta - \sec \theta) d\theta =$

$\displaystyle \frac{a^2}{2}(\sec \theta \tan \theta - \ln|\sec \theta + \tan \theta|) + C.$

$\displaystyle A = \frac{2b}{a}(\frac{a^2}{2})\,[\frac{x}{a}\frac{\sqrt{x^2 - a^2}}{a} - \ln\left|\frac{x}{a} + \frac{\sqrt{x^2 - a^2}}{a}\right|]\Big|_a^c$

$\displaystyle = ab[\frac{c\sqrt{c^2 - a^2}}{a^2} - \ln\left|\frac{c}{a} + \frac{\sqrt{c^2 - a^2}}{a}\right|]$

$\displaystyle = ab[\frac{cb}{a^2} - \ln(\frac{b+c}{a})]$ (Recall $c^2 = a^2 + b^2$).

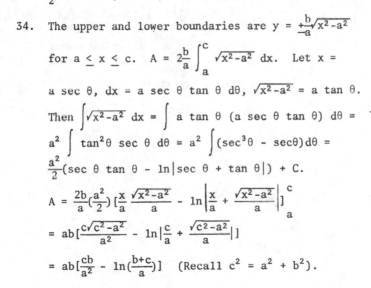

37. The one identity to remember is $\cosh^2 u = 1 + \sinh^2 u$. Let $x = 5 \sinh u$, $dx =$

$5 \cosh u$, $\sqrt{25 + x^2} = \sqrt{25(1 + \sinh^2 u)} = 5 \cosh u$. Then $\displaystyle\int \frac{1}{x^2\sqrt{25 + x^2}} \, dx =$

$\displaystyle\int \frac{5 \cosh u}{25 \sinh^2 u(5 \cosh u)} \, du = \frac{1}{25} \int \frac{1}{\sinh^2 u} \, du = \frac{1}{25} \int \operatorname{csch}^2 u \, du = -\frac{1}{25} \coth u + C$

$\displaystyle = -\frac{1}{25}\frac{\cosh u}{\sinh u} + C = -\frac{1}{25}\frac{\sqrt{1 + \sinh^2 u}}{\sinh u} + C = -\frac{1}{25}\frac{\sqrt{1 + x^2/25}}{x/5} + C = -\frac{1}{25}\frac{\sqrt{25 + x^2}}{x} + C.$

40. Let $x = 4 \tanh u$, $dx = 4 \operatorname{sech}^2 u \, du$, $16 - x^2 = 16(1 - \tanh^2 u) = 16 \operatorname{sech}^2 u$.

$\displaystyle\int \frac{1}{16 - x^2} \, dx = \int \frac{4 \operatorname{sech}^2 u}{16 \operatorname{sech}^2 u} \, du = \frac{1}{4} \int du = \frac{1}{4}u + C = \frac{1}{4} \tanh^{-1} \frac{x}{4} + C.$

EXERCISES 10.4, page 476

1. $\frac{5x-12}{x(x-4)} = \frac{A}{x} + \frac{B}{x-4}$. Multiplying by $x(x-4)$, we get $5x-12 = A(x-4) + Bx$. Letting $x = 4$ we find that $20-12 = A(0) + 4B$, or $8 = 4B$, or $B = 2$. Letting $x = 0$, we get $-12 = -4A$, or $A = 3$. Thus $\int \frac{5x-12}{x(x-4)} dx = \int (\frac{3}{x} + \frac{2}{x-4}) dx = 3 \ln|x| + 2 \ln|x-4| + C = \ln|x|^3 + \ln|x-4|^2 + C = \ln|x|^3|x-4|^2 + C$.

4. $\frac{4x^2+54x+134}{(x-1)(x+5)(x+3)} = \frac{A}{x-1} + \frac{B}{x+5} + \frac{C}{x+3}$. Multiplying by $(x-1)(x+5)(x+3)$, we get $4x^2+54x+134 = A(x+5)(x+3) + B(x-1)(x+3) + C(x-1)(x+5)$.

Set $x = 1$ to obtain $192 = 24A$,   or $A = 8$.
Set $x = -5$ to obtain $-36 = 12B$,   or $B = -3$.
Set $x = -3$ to obtain $8 = -8C$,   or $C = -1$.

$\int \frac{4x^2+54x+134}{(x-1)(x+5)(x+3)} dx = \int (\frac{8}{x-1} - \frac{3}{x+5} - \frac{1}{x+1}) dx = 8 \ln|x-1| - 3 \ln|x+5| - \ln|x+1| + C$, or $\ln \frac{|x-1|^8}{|x+5|^3|x+1|} + C$ where now C is the usual arbitrary constant.

7. $\frac{x+16}{x^2+2x-8} = \frac{x+16}{(x+4)(x-2)} = \frac{A}{x+4} + \frac{B}{x-2}$. Multiplying by $(x+4)(x-2)$ we get $x+16 = A(x-2) + B(x+4)$, and letting $x = 2$ we find $18 = 6B$ or $B = 3$, and letting $x = -4$, we find $12 = -6A$ or $A = -2$. Thus $\int \frac{x+16}{x^2+2x-8} dx = \int (\frac{-2}{x+4} + \frac{3}{x-2}) dx = -2 \ln|x+4| + 3 \ln|x-2| + C$.

10. $\frac{4x^2-5x-15}{x^3-4x^2-5x} = \frac{4x^2-5x-15}{x(x-5)(x+1)} = \frac{A}{x} + \frac{B}{x-5} + \frac{C}{x+1}$. $4x^2-5x-15 = A(x-5)(x+1) + Bx(x+1) + Cx(x-5)$.

Set $x = 0$ to obtain $-15 = -5A$ or $A = 3$.
Set $x = -1$ to obtain $-6 = 6C$ or $C = -1$.
Set $x = 5$ to obtain $60 = 30B$ or $B = 2$.

$\int \frac{4x^2-5x-15}{x^3-4x^2-5x} dx = \int (\frac{3}{x} + \frac{2}{x-5} - \frac{1}{x+3}) dx = 3 \ln|x| + 2 \ln|x-5| - \ln|x+3| + C = \ln \frac{|x|^3|x-5|^2}{|x+1|} + C$ where now C is arbitrary.

13. The integrand is $\frac{9x^4+17x^3+3x^2-8x+3}{x^4(x+3)} = \frac{A}{x} + \frac{B}{x^2} + \frac{C}{x^3} + \frac{D}{x^4} + \frac{E}{x+3}$.

$9x^4 + 17x^3 + 3x^2 - 8x + 3 = Ax^3(x+3) + Bx^2(x+3) + Cx(x+3) + D(x+3) + Ex^4$

Set $x = 0$ to obtain $3 = 3D$ or $D = 1$.

Set $x = -3$ to obtain $324 = 81E$ or $E = 4$.

To obtain the other constants, we set the coefficients of like powers of x equal to each other:

x^4: $9 = A + E = A + 4$, or $A = 5$

x^3: $17 = 3A + B = 15 + B$, or $B = 2$

x^2: $3 = 3B + C = 6 + C$, or $C = -3$.

$$\int \frac{9x^4 + 17x^3 + 3x^2 - 8x + 3}{x^4(x+3)}\,dx = \int \left(\frac{5}{x} + \frac{2}{x^2} - \frac{3}{x^3} + \frac{1}{x^4} + \frac{4}{x+3}\right)dx =$$

$$5\ln|x| - \frac{2}{x} + \frac{3}{2x^2} - \frac{1}{3x^3} + 4\ln|x+3| + C.$$

16. With $u = x-7$, $\displaystyle\int \frac{1}{(x-7)^5}\,dx = \int u^{-5}du = \frac{u^{-4}}{-4} + C = -1/4(x-7)^4 + C.$

19. $\dfrac{x^2+3x+1}{x^4+5x^2+4} = \dfrac{x^2+3x+1}{(x^2+4)(x^2+1)} = \dfrac{Ax+B}{x^2+4} + \dfrac{Cx+D}{x^2+1}$

$x^2+3x+1 = (Ax+B)(x^2+1) + (Cx+D)(x^2+4) = (A+C)x^3 + (B+D)x^2 + (A+4C)x + (B+4D)$

Equating coefficients of like powers of x, we obtain

x^3: $0 = A \quad\ + C$

x^2: $1 = \quad B \quad\quad + D$

x^1: $3 = A \quad +4C$

x^0: $1 = \quad B \quad\quad +4D$

Taking the 1st and 3rd, we get $A = -1$, $C = 1$, and from the 2nd and 4th, we get $B = 1$, $D = 0$. The integral of the original function reduces to

$$\int \left(\frac{-x}{x^2+4} + \frac{1}{x^2+4} + \frac{x}{x^2+1}\right)dx = -\frac{1}{2}\ln(x^2+4) + \frac{1}{2}\tan^{-1}\left(\frac{x}{2}\right) + \frac{1}{2}\ln(x^2+1) + C.$$

22. $\dfrac{x^4 + 2x^2 + 4x + 1}{(x^2+1)^3} = \dfrac{Ax+B}{x^2+1} + \dfrac{Cx+D}{(x^2+1)^2} + \dfrac{Ex+F}{(x^2+1)^3}$

$x^4+2x^2+4x+1 = (Ax+B)(x^2+1)^2 + (Cx+D)(x^2+1) + Ex + F$

$= Ax^5 + Bx^4 + (2A+C)x^3 + (2B+D)x^2 + (A+C+E)x + (B+D+F).$

Equating coefficients as above:

x^5: $0 = A$ x^4: $1 = B$

x^3: $0 = 2A + C$ x^2: $2 = 2B + D$

x^1: $4 = A + C + E$ x^0: $1 = B + D + F$

yielding $A = C = D = F = 0$, $B = 1$, $E = 4$. The integral reduces to

$$\int \left[\frac{1}{x^2+1} + \frac{4x}{(x^2+1)^3}\right]dx = \tan^{-1}x - \frac{1}{(x^2+1)^2} + C.$$

25. By long division $\dfrac{x^6-x^3+1}{x^4+9x^2} = x^2 - 9 + \dfrac{-x^3+81x^2+1}{x^4 + 9x^2}$. $\dfrac{-x^3+81x^2+1}{x^4+9x^2} = \dfrac{-x^3+81x^2+1}{x^2(x^2+9)} =$

$\dfrac{A}{x} + \dfrac{B}{x^2} + \dfrac{Cx+D}{x^2+9}$.

$-x^3+81x^2+1 = Ax(x^2+9) + B(x^2+9) + (Cx+D)x^2 = (A+C)x^3 + (B+D)x^2 + 9Ax + 9B.$

From x^0 and x^1 we get immediately $A = 0$, $B = 1/9$. From x^3, $-1 = A+C = C$;

from x^2, $81 = \frac{1}{9} + D$, or $D = 81 - \frac{1}{9} = \frac{728}{9}$. The integral reduces to

$$\int \left(x^2 - 9 + \frac{1}{9x^2} - \frac{x}{x^2+9} + \frac{728}{9}\cdot\frac{1}{x^2+9}\right)dx = \frac{x^3}{3} - 9x - \frac{1}{9x} - \frac{1}{2}\ln(x^2+9) + \frac{728}{27}\tan^{-1}\frac{x}{3} + C.$$

28. $\dfrac{-2x^4-3x^3-3x^2+3x+1}{x^2(x+1)^3} = \dfrac{A}{x} + \dfrac{B}{x^2} + \dfrac{C}{x+1} + \dfrac{D}{(x+1)^2} + \dfrac{E}{(x+1)^3}$.

$-2x^4 - 3x^3 - 3x^2 + 3x + 1 = Ax(x+1)^3 + B(x+1)^3 + Cx^2(x+1)^2 + Dx^2(x+1) + Ex^2$.

Set $x = 0$ to get $1 = B$, and set $x = -1$ to get $-4 = E$. To obtain the remaining constants, equate coefficients of three powers of x.

x^4: $-2 = A$ $+ C$

x^2: $-3 = 3A + 3B + C + D + E$

x : $3 = A + 3B$ $= A + 3 \Longrightarrow A = 0$

Using $A = 0$ in the first equation, we get $C = -2$. Using these in the second equation, we get $-3 = 0 + 3 - 2 + D - 4$, or $D = 0$. Thus the given integral reduces to $\displaystyle\int \ (\dfrac{1}{x^2} - \dfrac{2}{x+1} - \dfrac{4}{(x+1)^3})\,dx = -\dfrac{1}{x} - 2\ \ln|x+1| + \dfrac{2}{(x+1)^2} + C.$

31. Note that $x^3-x^2+x-1 = x^2(x-1) + (x-1) = (x-1)(x^2+1)$, and, by long division, the integrand can be written as $2x + (4x^2-3x+1)/(x-1)(x^2+1)$.

$\dfrac{4x^2-3x+1}{(x-1)(x^2+1)} = \dfrac{A}{x-1} + \dfrac{Bx+C}{x^2+1}$. $4x^2-3x+1 = A(x^2+1) + (Bx+C)(x-1)$.

Set $x = 1$ to obtain $2 = 2A$, or $A = 1$, and set $x = 0$ to obtain $1 = A-C = 1-C$. Thus $C = 0$. For the third constant we could equate some coefficient or simply pick another value of x. Let's set $x = -1$ to obtain $8 = 2A + 2B - 2C = 2 + 2B$, or $B = 3$. Thus the original integral reduces to $\displaystyle\int\ (2x + \dfrac{1}{x-1} + \dfrac{3x}{x^2+1})\,dx = x^2 + \ln|x-1| + \dfrac{3}{2}\ln(x^2+1) + C.$

34. $1/u(a + bu) = A/u + B/(a + bu) \Longrightarrow 1 = A(a + bu) + Bu$. Set $u = 0$ to obtain $1 = Aa$, or $A = 1/a$. Set $u = -a/b$ to obtain $1 = -Ba/b$, or $B = -b/a$. Thus

$\displaystyle\int\ \dfrac{1}{u(a+bu)}\ du = \int\ (\dfrac{1/a}{u} - \dfrac{b/a}{a+bu})\,du = (1/a)(\ln|u| - \ln|a+bu|) + C = \dfrac{1}{a}\ \ln\Big|\dfrac{u}{a+bu}\Big| + C.$

37. $f(x) = x/(x^2-2x-3) = x/(x-3)(x+1)$. Since $x-3 < 0$, $x \ge 0$, and $(x+1) > 0$ on $[0,2]$, it follows that $f(x) \le 0$ there. Thus, by (6.2), $A = -\displaystyle\int_0^2 f(x)\,dx.$

Now, $f(x) = \dfrac{x}{(x-3)(x+1)} = \dfrac{A}{x-3} + \dfrac{B}{x+1} \Longrightarrow x = A(x+1) + B(x-3)$. Setting $x = 3$ and -1, we obtain $3 = 4A$, or $A = 3/4$, and $-1 = -4B$, or $B = 1/4$. Thus $A =$

$-\displaystyle\int_0^2 (\dfrac{3/4}{x-3} + \dfrac{1/4}{x+1})\,dx = -[\dfrac{3}{4}\ \ln|x-3| + \dfrac{1}{4}\ \ln|x+1|]\Big|_0^2 = -\dfrac{1}{4}[(3\ \ln|-1| + \ln\ 3) - $

$(3\ \ln|-3| + \ln\ 1)] = -\dfrac{1}{4}(0 + \ln\ 3 - 3\ \ln\ 3 + 0) = \dfrac{2\ \ln\ 3}{4}$.

40. The distance travelled in the time interval $[1,2]$ is $s(2) - s(1) = \displaystyle\int_1^2 v(t)\,dt$

$= \displaystyle\int_1^2\ (t+3)/(t^3+t)\,dt.$ Now, $\dfrac{t+3}{t^3+t} = \dfrac{t+3}{t(t^2+1)} = \dfrac{A}{t} + \dfrac{Bt+C}{t^2+1} \Longrightarrow t+3 = A(t^2+1) + $

$(Bt+C)t$ or $t+3 = (A+B)t^2 + Ct + A$. Thus $A = 3$, $C = 1$ and $A+B = 0$, or $B =$

$-A = -3$. Thus $s(2) - s(1) = \int_1^2 (\frac{3}{t} - \frac{3t}{t^2+1} + \frac{1}{t^2+1})dt = 3\ln|t|]_1^2 - \frac{3}{2}\ln(t^2+1)]_1^2 +$

$\tan^{-1}t]_1^2 = 3\ln 2 - (\frac{3}{2}\ln 5 - \frac{3}{2}\ln 2) + \tan^{-1}2 - \tan^{-1}1 = (9/2)\ln 2 -$

$(3/2)\ln 5 + \tan^{-1}2 - \pi/4 \approx 1.027$.

EXERCISES 10.5, page 480

1. $x^2-4x+8 = (x-2)^2 + 4$. So, with $u = x-2$, $du = dx$, $\int \frac{dx}{x^2-4x+8} = \int \frac{dx}{(x-2)^2+4}$

$= \int \frac{du}{u^2+4} = \frac{1}{2}\tan^{-1}\frac{u}{2} + C = \frac{1}{2}\tan^{-1}\frac{x-2}{2} + C$. Alternately, after completing the

square, we could have let $x-2 = 2\tan\theta$, $dx = 2\sec^2\theta\,d\theta$ and $(x-2)^2 + 4 =$

$4\sec^2\theta$. Note also that $\tan\theta = (x-2)/2 \Rightarrow \theta = \tan^{-1}((x-2)/2)$. The integral

then becomes $\int \frac{2\sec^2\theta}{4\sec^2\theta}d\theta = \frac{1}{2}\int d\theta = \frac{1}{2}\theta + C = \frac{1}{2}\tan^{-1}\frac{x-2}{2} + C$ as before. Both

methods are equivalent. Use whichever you prefer.

2. $7 + 6x - x^2 = 16 - (x-3)^2$. So, with $u = x-3$, $du = dx$. $\int \frac{dx}{\sqrt{7+6x-x^2}} = \int \frac{du}{\sqrt{16-u^2}}$

$= \sin^{-1}\frac{u}{4} + C = \sin^{-1}\frac{x-3}{4} + C$.

4. Proceed as in #1 using $x^2-2x+2 = (x-1)^2 + 1$ to get $\tan^{-1}(x-1) + C$.

7. $(x^3-1) = (x-1)(x^2+x+1) = (x-1)[(x+\frac{1}{2})^2 + \frac{3}{4}]$. Let $u = x + \frac{1}{2}$ so that $x-1 =$

$u - \frac{3}{2}$ and $\frac{1}{x^3-1} = \frac{1}{(u-\frac{3}{2})(u^2+\frac{3}{4})} = \frac{A}{u-\frac{3}{2}} + \frac{Bu+C}{u^2+\frac{3}{4}}$. As in the previous section,

this yields $1 = A(u^2 + \frac{3}{4}) + (Bu + C)(u - \frac{3}{2})$.

Set $u = 3/2$ to obtain $1 = 3A$ or $A = 1/3$.

Set $u = 0$ to obtain $1 = \frac{3}{4}A - \frac{3}{2}C$ or $C = -1/2$.

Equate u^2 coefficients: $0 = A+B$, or $B = -1/3$. Now we can finally tie this

all together:

$\int \frac{1}{x^3-1}dx = \int [\frac{1/3}{u-3/2} - \frac{1/3\,u}{u^2+3/4} - \frac{1/2}{u^2+3/4}]du = \frac{1}{3}\ln|u - \frac{3}{2}| - \frac{1}{6}\ln(u^2+\frac{3}{4})$

$- \frac{1}{\sqrt{3}}\tan^{-1}\frac{2u}{\sqrt{3}} + C = \frac{1}{3}\ln|x-1| - \frac{1}{6}\ln(x^2+x+1) - \frac{1}{\sqrt{3}}\tan^{-1}\frac{2x+1}{\sqrt{3}} + C$.

10. $x^4 - 4x^3 + 13x^2 = x^2(x^2-4x+13) = x^2[(x-2)^2 + 9]$. Let $u = x-2$ so that $x =$

$u+2$ and $\frac{1}{x^4-4x^3+13x^2} = \frac{1}{(u+2)^2(u^2+9)} = \frac{A}{u+2} + \frac{B}{(u+2)^2} + \frac{Cu+D}{u^2+9}$. Proceeding as in

the previous section, the solution is $A = 4/169$, $B = 1/13$, $C = -4/169$,

$D = -5/169$, and $\int \frac{1}{x^4-4x^3+13x^2}dx = \frac{4}{169}\ln|u+2| - \frac{1}{13}\cdot\frac{1}{u+2} - \frac{2}{169}\ln(u^2+9)$

$$- \frac{5}{3(169)} \tan^{-1} \frac{u}{3} + C = \frac{4}{169} \ln|x| - \frac{1}{13x} - \frac{2}{169} \ln(x^2-4x+13) - \frac{5}{507} \tan^{-1} \frac{x-2}{3} + C.$$

13. $2x^2 - 3x + 9 = 2(x^2 - \frac{3}{2}x + \frac{9}{16} + (\frac{9}{2} - \frac{9}{16})) = 2((x-3/4)^2 + 63/16)$. So, with

$u = x - \frac{3}{4}$, $\int \frac{1}{2x^2-3x+9} dx = \frac{1}{2} \int \frac{1}{(x-3/4)^2+63/16} dx = \frac{1}{2} \int \frac{1}{u^2+(3\sqrt{7}/4)^2} du =$

$\frac{1}{2} \cdot \frac{4}{3\sqrt{7}} \tan^{-1} \frac{4u}{3\sqrt{7}} + C = \frac{2}{3\sqrt{7}} \tan^{-1} \frac{4(x-3/4)}{3\sqrt{7}} + C = \frac{2}{3\sqrt{7}} \tan^{-1} \frac{(4x-3)}{3\sqrt{7}} + C.$

16. By the disc method, $V = \pi \int_0^2 1/(x^2+2x+10)^2 dx = \pi \int_1^2 1/[(x+1)^2+9]^2 dx$. Let $u =$

x+1, du = dx. When x = 0 or 2, then u = 1 or 3, and $V = \pi \int_1^3 1/(u^2+9)^2 du.$

Let $u = 3 \tan \theta$, $du = 3 \sec^2\theta\, d\theta$, $u^2+9 = 9 \sec^2\theta$. Then $\int \frac{1}{(u^2+9)^2} du =$

$\int \frac{3 \sec^2\theta}{81 \sec^4\theta} d\theta = \frac{1}{27} \int \frac{1}{\sec^2\theta} d\theta = \frac{1}{27} \int \cos^2\theta\, d\theta = \frac{1}{54} \int (1 + \cos 2\theta) d\theta =$

$\frac{1}{54}(\theta + \frac{1}{2} \sin 2\theta) = \frac{1}{54}(\theta + \sin \theta \cos \theta) =$

$\frac{1}{54}(\tan^{-1} \frac{u}{3} + \frac{3u}{u^2+9})$. Thus

$V = \frac{1}{54}[(\tan^{-1} \frac{3}{3} + \frac{9}{18}) - (\tan^{-1} \frac{1}{3} + \frac{3}{10})] =$

$\frac{1}{54}(\frac{\pi}{4} + \frac{1}{5} - \tan^{-1} \frac{1}{3}) \approx 0.0386.$

EXERCISES 10.6, page 484

1. Let $u = \sqrt[3]{x+9}$. Then $u^3 = x+9$, $x = u^3-9$, $dx = 3 u^2 du$. $\int x\sqrt[3]{x+9}\, dx =$

$\int (u^3-9)u(3u^2) du = 3 \int (u^6-9u^3) du = 3(\frac{u^7}{7} - \frac{9}{4}u^4) + C$, where $u = (x+9)^{1/3}$.

4. Let $u = (x+3)^{1/3}$. Then $u^3 = x+3$, $x = u^3-3$, $dx = 3 u^2 du$. $\int \frac{5x}{(x+3)^{2/3}} dx$

$= 5 \int \frac{(u^3-3)(3u^2)}{u^2} du = 15 \int (u^3-3) du = 15(\frac{u^4}{4} - 3u) + C$

$= 15(\frac{(x+3)^{4/3}}{4} - 3(x+3)^{1/3}) + C.$

7. As in Example 2, let $x = u^6$. Then $dx = 6u^5 du$, $\sqrt{x} = u^3$, $\sqrt[3]{x} = u^2$. $\int \frac{\sqrt{x}}{1+\sqrt[3]{x}} dx$

$= \int \frac{u^3(6u^5)}{1+u^2} du = 6 \int \frac{u^8}{1+u^2} du = 6 \int (u^6-u^4+u^2-1+\frac{1}{1+u^2}) du$ (by long division) $=$

$6[\frac{u^7}{7} - \frac{u^5}{5} + \frac{u^3}{3} - u + \tan^{-1}u] + C = 6[\frac{x^{7/6}}{7} - \frac{x^{5/6}}{5} + \frac{x^{1/2}}{3} - x^{1/6} + \tan^{-1}(x^{1/6})]+C.$

10. Let $u = \sqrt{1+2x}$. Then $u^2 = 1+2x$, $2u\, du = 2 dx$ or $dx = u\, du$, and $2x+3 =$

$(u^2-1)+3 = u^2+2.$ Then $\displaystyle\int \frac{2x+3}{\sqrt{1+2x}}\, dx = \int \frac{(u^2+2)u\, du}{u} = \frac{u^3}{3} + 2u + C = \frac{(1+2x)^{3/2}}{3}$

$+ 2(1+2x)^{1/2} + C = \dfrac{2x+7}{3}\sqrt{1+2x} + C.$

13. Let $u = 1 + e^x$. Then $du = e^x dx$, $e^x = u-1$, and $e^{2x} = (u-1)^2$. $\displaystyle\int e^{3x}\sqrt{e^x+1}\, dx$

$= \displaystyle\int e^{2x}\sqrt{e^x+1}\, e^x dx = \int (u-1)^2\, u^{1/2}\, du = \int (u^2-2u+1)u^{1/2}\, du =$

$\displaystyle\int (u^{5/2} - 2u^{3/2} + u^{1/2})\, du = \frac{2}{7}u^{7/2} - \frac{4}{5}u^{5/2} + \frac{2}{3}u^{3/2} + C$, where $u = (1+e^x)$.

16. $\displaystyle\int \frac{\sin 2x}{\sqrt{1+\sin x}}\, dx = 2\int \frac{\sin x \cos x}{\sqrt{1+\sin x}}\, dx$. Let $u = 1+\sin x$. Then $du = \cos x\, dx$,

$\sin x = u-1$, and the integral becomes $2\displaystyle\int \frac{u-1}{u^{1/2}}\, du = 2\int (u^{1/2} - u^{-1/2})\, du =$

$2(\frac{2}{3}u^{3/2} - 2u^{1/2}) + C = 4(\dfrac{(1+\sin x)^{3/2}}{3}) - (1+\sin x)^{1/2}) + C.$

19. Let $u = x-1$. Then $du = dx$ and $x = u+1$. $\displaystyle\int \frac{x}{(x-1)^6}\, dx = \int \frac{u+1}{u^6}\, du =$

$\displaystyle\int (u^{-5} + u^{-6})\, du = \frac{u^{-4}}{-4} + \frac{u^{-5}}{-5} + C = -\frac{1}{4(x-1)^4} - \frac{1}{5(x-1)^5} + C.$

22. Using $z = \tan(x/2)$, $\cos x = (1-z^2)/(1+z^2)$, $dx = \dfrac{2}{1+z^2}\, dz$, $\displaystyle\int \frac{1}{3 + 2\cos x}\, dx$

$= \displaystyle\int \frac{2/(1+z^2)}{3 + \frac{2(1-z^2)}{1+z^2}}\, dz = \int \frac{2\, dx}{[3(1+z^2) + 2(1-z^2)]} = 2\int \frac{dz}{5+z^2} = \frac{2}{\sqrt 5} \tan^{-1}\frac{z}{\sqrt 5} + C.$

$= \dfrac{2}{\sqrt 5} \tan^{-1}(\dfrac{\tan(x/2)}{\sqrt 5}) + C.$

25. $I = \displaystyle\int \frac{\sec x}{4 - 3\tan x}\, dx = \int \frac{1/\cos x}{4 - 3\sin x/\cos x}\, dx = \int \frac{1}{4\cos x - 3\sin x}\, dx.$

As above, let $z = \tan x/2$, etc. $I = \displaystyle\int \frac{1}{4(\frac{1-z^2}{1+z^2}) - 3(\frac{2z}{1+z^2})} \cdot \frac{2}{1+z^2}\, dz$

$= \displaystyle\int \frac{2}{4(1-z^2) - 6z}\, dz = \int \frac{-1}{2z^2+3z-2}\, dz.$ Now, as in Section 10.4, $\dfrac{-1}{2z^2+3z-2} =$

$\dfrac{-1}{(2z-1)(z+2)} = \dfrac{A}{2z-1} + \dfrac{B}{z+2} \implies -1 = A(z+2) + B(2z-1).$ Setting $z = -2$ and $1/2$

we get $-1 = -5B$, or $B = 1/5$, and $-1 = (5/2)A$, or $A = -2/5$. Thus $I =$

$\displaystyle\int (\frac{-2/5}{2z-1} + \frac{1/5}{z+1})\, dz = -\frac{1}{5}\ln|2z-1| + \frac{1}{5}\ln|z+1| + C = -\frac{1}{5}\ln|2\tan\frac{x}{2} - 1| +$

$\frac{1}{5}\ln|\tan\frac{x}{2} + 1| + C.$

EXERCISES 10.7, page 486

1. The presence of $\sqrt{4+9x^2}$ refers us to the portion of the table involving
    $\sqrt{a^2+u^2}$. The integrand, $\sqrt{4+9x^2}/x$ indicates that Formula 23 is the one to use.
    To convert the given integral into the desired form, we see that $a^2 = 4$,

$u^2 = 9x^2$ so that a = 2, u = 3x, du = 3 dx. If we multiply and divide by 3, then u = 3x is in the denominator, and du = 3 dx is in the numerator. Then

$$\int \frac{\sqrt{4+9x^2}}{x} \, dx = \int \frac{\sqrt{2^2 + (3x)^2}}{3x} \, 3 \, dx = \sqrt{2^2 + (3x)^2} - 2 \ln\left|\frac{2 + \sqrt{2^2 + (3x)^2}}{3x}\right| + C,$$

which quickly reduces to the given answer.

4. The integrand, $x^2\sqrt{4x^2-16}$, indicates that Formula 40 is the one to use. We can make the given integral compatible with that formula as above, or, alternately, by changing variables. Let u = 2x, x = u/2, dx = du/2. Then

$$\int x^2\sqrt{4x^2-16} \, dx = \int \frac{u^2}{4} \sqrt{u^2-16} \cdot \frac{1}{2} \, du = \frac{1}{8} \int u^2\sqrt{u^2-16} \, du,$$ which is in the correct

form with a = 4. Using Formula 40, we obtain $\frac{1}{8}[\frac{u}{8}(2u^2-16)\sqrt{u^2-16} - \frac{4^4}{8} \cdot$

$\ln|u + \sqrt{u^2-16}| + C = \frac{1}{8}[\frac{2x}{8}(2(4x^2)-16)\sqrt{4x^2-16} - \frac{256}{8} \ln|2x + \sqrt{4x^2-16}| + C$

$= \frac{x}{4}(x^2-2)\sqrt{4x^2-16} - 4 \ln|2x - \sqrt{4x^2-16}| + C.$

7. Let u = 3x, dx = (1/3) du. Then by Formula 73 with n = 6, $I = \frac{1}{3}\int \sin^6 u \, du =$

$\frac{1}{3}[-\frac{1}{6} \sin^5 u \cos u + \frac{5}{6} \int \sin^4 u \, du].$ By 73 again, with n = 4, $\int \sin^4 u \, du =$

$-\frac{1}{4} \sin^3 u \cos u + \frac{3}{4} \int \sin^2 u \, du.$ Using Formula 63 for the last integral, we

obtain: $I = \frac{1}{3}[-\frac{1}{6} \sin^5 u \cos u - \frac{5}{24} \sin^3 u \cos u + \frac{15}{24}(\frac{u}{2} - \frac{\sin 2u}{4})] + C,$ where

u = 3x. (sin 2u can be replaced by 2 sin u cos u if desired.)

10. Directly from Formula 81 with a = 5, b = 3, and u = x, $\int \sin 5x \cos 3x \, dx =$

$-\frac{\cos 2x}{2(2)} - \frac{\cos 8x}{2(8)} + C.$

13. Directly from Formula 98 with a = -3, b = 2, and u = x, $\int e^{-3x} \sin 2x \, dx =$

$\frac{e^{-3x}}{(-3)^2+2^2} (-3 \sin 2x - 2 \cos 2x) + C.$

16. From the integrand, $1/x\sqrt{3x-2x^2}$, the formulas involving $\sqrt{2au-u^2}$ should be consulted, and we see that Formula 120 fits. So, let $u = \sqrt{2}x$ so that $u^2 = 2x^2$,

x = u/\sqrt{2}, dx = du/\sqrt{2}. Then $I = \int \frac{dx}{x\sqrt{3x-2x^2}} = \int \frac{du/\sqrt{2}}{(u/\sqrt{2})\sqrt{3(u/\sqrt{2})-u^2}} =$

$\int \frac{1}{u\sqrt{\frac{3}{\sqrt{2}}u - u^2}} \, du.$ Thus $2a = \frac{3}{\sqrt{2}},$ or $a = \frac{3}{2\sqrt{2}}.$ By the formula,

$I = -\frac{\sqrt{(3/\sqrt{2})u-u^2}}{(3/2\sqrt{2})u} + C = -\frac{\sqrt{3x-2x^2}}{(3/2)x} + C.$

19. Writing $\int e^{2x} \cos^{-1} e^x \, dx = \int e^x \cos^{-1} e^x (e^x) \, dx,$ we let $u = e^x,$ $du = e^x dx$ to

$\int u \cos^{-1}u \, du$, which, by Formula 91, equals $\dfrac{2u^2-1}{4}\cos^{-1}u - \dfrac{u\sqrt{1-u^2}}{4} + C$, where $u = e^x$.

EXERCISES 10.8, page 495

NOTE: As in Chapter 6, in addition to the solutions of #1, 4, 7, ..., I have included the integrals from which the answers to the remaining odd-numbered problems
can be obtained in this section and in the next section as well. In these answers,
the density, ρ, is taken as unity.

1. $m = 2 + 7 + 5 = 14$. $M_x = \sum\limits_{i=1}^{3} m_i y_i = 2(-1) + 7(0) + 5(-5) = -27$.

$M_y = \sum\limits_{i=1}^{3} m_i x_i = 2(4) + 7(-2) + 5(-8) = -46$.

$\bar{x} = \dfrac{M_y}{m} = -\dfrac{46}{14}$, $\bar{y} = \dfrac{M_x}{m} = -\dfrac{27}{14}$.

3. $m = \int_0^1 x^3 dx$, $M_x = (1/2)\int_0^1 x^6 dx$, $M_y = \int_0^1 x^4 dx$.

4. With $\rho = 1$, $m = $ Area $= \int_0^9 x^{1/2} dx = \dfrac{2}{3}x^{3/2} \Big]_0^9 = 18$.

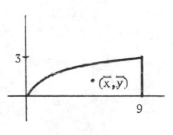

$M_y = \int_0^9 x \cdot x^{1/2} dx = \dfrac{2}{5}x^{5/2}\Big]_0^9 = \dfrac{486}{5}$. $M_x = $

$\int_0^9 \dfrac{1}{2}(x^{1/2})^2 dx = \dfrac{x^2}{4}\Big]_0^9 = \dfrac{81}{4}$, $\bar{x} = \dfrac{486}{5(18)} = \dfrac{27}{5}$,

$\bar{y} = \dfrac{81}{4(18)} = \dfrac{9}{8}$.

5. $m = \int_0^\pi \sin x \, dx$, $M_x = (1/2)\int_0^\pi \sin^2 x \, dx$, $M_y = \int_0^\pi x \sin x \, dx$.

7. $m = A = \int_{-2}^1 [(1-x^2) - (x-1)] dx = \int_{-2}^1 (2-x-x^2) dx = [2x - \dfrac{x^2}{2} - \dfrac{x^3}{3}]_{-2}^1 = \dfrac{9}{2}$.

$M_y = \int_{-2}^1 x(2-x-x^2) dx = [x^2 - \dfrac{x^3}{3} - \dfrac{x^4}{4}]_{-2}^1 = -\dfrac{9}{4}$. $M_x = \dfrac{1}{2}\int_{-2}^1 [(1-x^2)^2-(x-1)^2] dx$

$= \dfrac{1}{2}\int_{-2}^1 (x^4-3x^2+2x) dx = \dfrac{1}{2}[\dfrac{x^5}{5} - x^3 + x^2]_{-2}^1 = \dfrac{1}{2}(-\dfrac{27}{5}) = -\dfrac{27}{10}$.

$\bar{x} = -\dfrac{9}{4} \cdot \dfrac{2}{9} = -\dfrac{1}{2}$, $\bar{y} = -\dfrac{27}{10} \cdot \dfrac{2}{9} = -\dfrac{3}{5}$.

9. $m = \int_0^3 1/\sqrt{16+x^2}\ dx$, $M_x = (1/2) \int_1^3 1/(16+x^2)\,dx$, $M_y = \int_0^3 x/\sqrt{16+x^2}\ dx$.

10. $m = \int_1^e \frac{1}{x}\ dx = \ln x\ ]_1^e = 1$. $M_y = \int_1^e x \cdot \frac{1}{x}\ dx = e-1$. $M_x = \frac{1}{2} \int_1^e (\frac{1}{x})^2 dx = $

   $\frac{1}{2}[- \frac{1}{x}]_1^e = \frac{1}{2}(1 - \frac{1}{e})$. $\bar{x} = e-1$, $\bar{y} = \frac{1}{2}(1 - \frac{1}{e})$.

11. $m = \int_{-1}^0 e^{2x}dx$, $M_x = (1/2) \int_{-1}^0 e^{4x}dx$, $M_y = \int_{-1}^0 xe^{2x}dx$.

13. The 3 medians have equations $y = \frac{b+c}{a} x$, $y = $
   $\frac{2b-c}{2a} x + \frac{c}{2}$, and $y = \frac{b-2c}{a} x + c$, and their point
   of intersection is $(\frac{a}{3},\frac{b+c}{3})$. The area is $m = \frac{1}{2}ac$
   (without integrating). The upper boundary of the
   region is $f(x) = \frac{b-c}{a} x + c$; the lower boundary

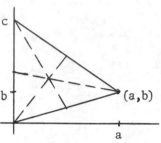

   $g(x) = \frac{b}{a} x$. $M_y = \int_0^a x(\frac{b-c}{a} x + c - \frac{b}{a} x)dx = c \int_0^a (x - \frac{x^2}{a})dx = \frac{ca^2}{6}$.

   $M_x = \frac{1}{2} \int_0^a [(\frac{b-c}{a} x + c)^2 - (\frac{b}{a} x)^2]dx = \frac{c}{2} \int_0^a (\frac{c-2b}{a^2} x^2 + \frac{2(b-c)}{a} x + c)dx$

   $= \frac{ac}{6}(b+c)$. $\bar{x} = \frac{a^2 c}{6} \cdot \frac{2}{ac} =, \frac{a}{3}$. $\bar{y} = \frac{ac}{6}(b+c) \cdot \frac{2}{ac} = \frac{b+c}{3}$.

15. $m = \int_{-a}^a \sqrt{a^2-x^2}\ dx$ (but there's an obvious way to get m without integration),
   $M_x = (1/2) \int_{-a}^a (a^2-x^2)dx$, $M_y = \int_{-a}^a x\sqrt{a^2-x^2}\ dx$.

16. Because the region is symmetric about the y-axis,
   $\bar{x} = 0$. $m = \int_{-2p}^{2p} (p - \frac{x^2}{4p})dx = \frac{8p^2}{3}$.

   $M_x = \frac{1}{2} \int_{-2p}^{2p} [p^2 - (\frac{x^2}{4p})^2]dx = \frac{8}{5}p^3$.

   $\bar{y} = \frac{8p^3}{5} \cdot \frac{3}{8p^2} = \frac{3}{5}p$.

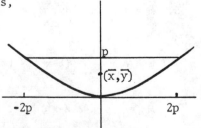

17. m = (area of square of side 2a) + (area of circle of radius a). With x-axis
   placed on the top edge of the square, origin at the center (also the center
   of the circle), $M_x = -(1/2) \int_{-a}^a (-2a)^2 dx + (1/2) \int_{-a}^a (a^2-x^2)dx$, $M_y = 0$ by
   symmetry.

19. $\bar{x} = 0$ by symmetry and $m = \pi ab/2$ by Exercise 33, Sec. 10.3 or Example 4, Sec.

7.3. $M_x = \dfrac{1}{2}\displaystyle\int_{-a}^{a} \dfrac{b^2}{a^2}\,(a^2-x^2)\,dx = \dfrac{2ab^2}{3}$. $\bar{y} = \dfrac{2ab^2}{3}\cdot\dfrac{2}{\pi ab} = \dfrac{4b}{3\pi}$.

EXERCISES 10.9, page 501

NOTE: As in Sec. 10.8, $\rho = 1$ and m is the volume of the solid.

1. $m = \pi\displaystyle\int_{1}^{2} (\tfrac{1}{x})^2\,dx = \pi[-\tfrac{1}{x}]\,\Big]_{1}^{2} = \dfrac{\pi}{2}$.

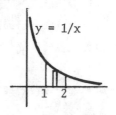

$M_{yz} = \pi\displaystyle\int_{1}^{2} x(\tfrac{1}{x})^2\,dx = \pi \ln x\,]\,\Big]_{1}^{2} = \pi \ln 2.$

$\bar{x} = \dfrac{\pi \ln 2}{\pi/2} = 2 \ln 2 \approx 1.386.$

3. $m = \pi\displaystyle\int_{0}^{1} e^{2x}\,dx$, $M_{yz} = \pi\displaystyle\int_{0}^{1} xe^{2x}\,dx.$

4. $m = \pi\displaystyle\int_{0}^{4} (\dfrac{1}{\sqrt{16+x^2}})^2\,dx = \dfrac{\pi}{4} \tan^{-1} \dfrac{x}{4}\,]\,\Big]_{0}^{4} = (\dfrac{\pi}{4})^2.$

$M_{yz} = \pi\displaystyle\int_{0}^{4} \dfrac{x}{16+x^2}\,dx = \dfrac{\pi}{2} \ln(16+x^2)]\,\Big]_{0}^{4} = \dfrac{\pi}{2}(\ln 32 - \ln 16) = \dfrac{\pi}{2} \ln 2.$

$\bar{x} = \dfrac{\pi \ln 2}{2}\cdot\dfrac{16}{\pi^2} = \dfrac{8 \ln 2}{\pi} \approx 1.77.$

5. $m = \pi\displaystyle\int_{0}^{1} (x^{2/3}-x^4)\,dx$, $M_{yz} = \pi\displaystyle\int_{0}^{1} (x^{5/3}-x^5)\,dx.$

7. It is convenient to use y as the integration
variable. $m = \pi\displaystyle\int_{2}^{4} (\sqrt{y^2-4})^2\,dy = \dfrac{32\pi}{3}$.

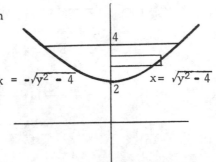

$M_{xz} = \pi\displaystyle\int_{2}^{4} y(y^2-4)\,dy = 36\pi.$

$\bar{y} = 36\pi\cdot\dfrac{3}{32\pi} = \dfrac{27}{8} = 3.375.$

9. $m = 2\pi\displaystyle\int_{0}^{2} xe^{2x}\,dx$, $M_{xz} = \pi\displaystyle\int_{0}^{2} xe^{4x}\,dx.$

10. Using the shell method as in Example 3, $m = 2\pi\displaystyle\int_{-1}^{0} (-x)e^{-x}\,dx = 2\pi[(x+1)e^{-x}\,]\,\Big]_{-1}^{0}$

$= 2\pi$. (Since $-1 < x < 0$, the radius of the shell is not w_i but $-w_i$. Hence the factor $(-x)$ in the integral.) The moment of the shell about the xz plane is $\frac{1}{2}$(height)$\cdot$(volume) or $\frac{1}{2}e^{-w_i}(2\pi(-w_i)e^{-w_i})\Delta x_i$. As in Chapter 6, we sum and

pass to the limit to obtain $M_{xz} = \pi \int_{-1}^{0} (-x)e^{-2x}dx = \frac{\pi}{4}(2x+1)e^{-2x} \Big]_{-1}^{0} = \frac{\pi}{4}(1+e^2)$.

$\bar{y} = \frac{\pi(1+e^2)}{4(2\pi)} = \frac{1+e^2}{8} \approx 1.05$.

(Note: the integrals for m and M_{xz} can both be done by integration by parts with $u = x$, $dv = e^{-x}dx$ or $e^{-2x}dx$, or use Formula 96 in the table of integrals.)

11. $m = 2\pi \int_{0}^{5} x/(x^2+25)dx$, $M_{xz} = \pi \int_{0}^{5} x/(x^2+25)^2dx$.

13. Using the disc method and (10.22) we revolve the right half of the region about the y-axis. The right boundary is $x = \frac{a}{b}\sqrt{b^2-y^2} = g(y)$. $m = \pi \int_{0}^{b} g(y)^2dy$

$= \frac{\pi a^2}{b^2} \int_{0}^{b} (b^2-y^2)dy = \frac{2\pi}{3}a^2b$. $M_{xz} = \pi \frac{a^2}{b^2} \int_{0}^{b} y(b^2-y^2)dy = \frac{\pi}{4} a^2b^2$.

$\bar{y} = \frac{\pi a^2b^2}{4} \cdot \frac{3}{2\pi a^2b} = \frac{3}{8} b$.

15. $m = 2\pi \int_{0}^{\pi/2} x \cos x \, dx$, $M_{xz} = \pi \int_{0}^{\pi/2} x \cos^2x \, dx$ (as in Example 3).

16. $m = \pi \int_{0}^{\pi/2} \cos^2x \, dx = \frac{\pi}{2} \int_{0}^{\pi/2} (1 + \cos 2x)dx = (\frac{\pi}{2})^2$. By (10.20),

$M_{yz} = \pi \int_{0}^{\pi/2} x \cos^2x \, dx$. Let $u = x$, $dv = \cos^2x \, dx$. Then $du = dx$,

$v = \frac{1}{2}(x + \frac{\sin 2x}{2})$ using $\cos^2x = \frac{1}{2}(1 + \cos 2x)$. Then $M_{yz} = \pi[\frac{x}{2}(x + \frac{\sin 2x}{2}) \Big]_{0}^{\pi/2}$

$- \frac{\pi}{2} \int_{0}^{\pi/2} (x + \frac{\sin 2x}{2})dx = \frac{\pi^3}{8} - \frac{\pi}{2}[\frac{x^2}{2} - \frac{\cos 2x}{4} \Big]_{0}^{\pi/2} = \frac{\pi^3}{8} - \frac{\pi}{2}(\frac{\pi^2}{8} + \frac{1}{2})$

$= \frac{\pi}{16}(\pi^2 - 4)$. $\bar{x} = \frac{\pi(\pi^2-4)}{16} \cdot \frac{4}{\pi^2} = \frac{\pi^2-4}{4\pi} \approx 0.467$.

19. The region is a quarter circle of radius a. Its area is $A = (1/4)\pi a^2$. When the region is revolved about either axis, a hemisphere of radius a is swept out whose volume is $V = \frac{1}{2} \cdot \frac{4}{3} \pi a^3 = \frac{2}{3} \pi a^3$. If the centroid is $(\bar{x},\bar{y})$, then by the formula near the top of p. 501, $V = 2\pi\bar{x}A$ and similarly $V = 2\pi\bar{y}A$. (In the first, V is the volume of the solid obtained by a revolution about the y-axis; in the second, revolution has been about the x-axis. Here, the same value of

V can be used on both.) Thus $\bar{x} = \bar{y} = \dfrac{V}{2\pi A} = \dfrac{(2/3)\pi a^3}{2\pi(1/4)\pi a^2} = \dfrac{4}{3\pi} a \approx 0.42\ a.$

EXERCISES 10.10 (Review), page 502

1. $u = \sin^{-1}x$, $dv = x\,dx \implies du = \dfrac{dx}{\sqrt{1-x^2}}$, $v = \dfrac{x^2}{2}$.

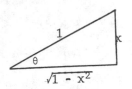

$I = \displaystyle\int x \sin^{-1}x\,dx = \dfrac{x^2}{2}\sin^{-1}x - \dfrac{1}{2}\int \dfrac{x^2}{\sqrt{1-x^2}}\,dx.$ With $x =$

$\sin\theta$, $dx = \cos\theta\,d\theta$, $\sqrt{1-x^2} = \cos\theta$, the last integral

is $\displaystyle\int \dfrac{\sin^2\theta\,\cos\theta}{\cos\theta}\,d\theta = \dfrac{1}{2}\int(1-\cos 2\theta)\,d\theta = \dfrac{1}{2}(\theta - \dfrac{\sin 2\theta}{2})$

$= \dfrac{1}{2}(\theta - \sin\theta\cos\theta) = \dfrac{1}{2}(\sin^{-1}x - x\sqrt{1-x^2}).$ Thus

$I = \dfrac{x^2}{2}\sin^{-1}x - \dfrac{1}{4}\sin^{-1}x + \dfrac{1}{4}x\sqrt{1-x^2} + C.$

4.   Let $y^2 = x$, $2y\,dy = dx$, $y = \sqrt{x}$. Then $\displaystyle\int_0^1 e^{\sqrt{x}}\,dx = 2\int_0^1 y\,e^y\,dy = 2(y-1)e^y\ ]_0^1$

$= 2.$ (The y integral was done by parts with $u = y$, $dv = e^y\,dy$.)

7. $\displaystyle\int \tan x\,\sec^5 x\,dx = \int \sec^4 x(\sec x\,\tan x)\,dx = \dfrac{\sec^5 x}{5} + C.$

10.   Let $x = 4\sin\theta$, $dx = 4\cos\theta\,d\theta$, $\sqrt{16-x^2} = 4\cos\theta$. Then $\displaystyle\int \dfrac{1}{x^2\sqrt{16-x^2}}\,dx =$

$\displaystyle\int \dfrac{4\cos\theta\,d\theta}{(16\sin^2\theta)(4\cos\theta)} = \dfrac{1}{16}\int \csc^2\theta\,d\theta = -\dfrac{1}{16}\cot\theta + C$

$= -\dfrac{1}{16}\dfrac{\sqrt{16-x^2}}{x} + C.$

13. $\dfrac{x^3+1}{x(x-1)^3} = \dfrac{A}{x} + \dfrac{B}{x-1} + \dfrac{C}{(x-1)^2} + \dfrac{D}{(x-1)^3}.$ $x^3+1 = A(x-1)^3 + Bx(x-1)^2 + Cx(x-1) +$

Dx. Setting $x = 1$ and then $x = 0$, we see $D = 2$, $A = -1$. Equating the x^3

coefficients: $1 = A+B = -1 + B \implies B = 2$; and the x^2 coefficients: $0 = -3A$

$-2B + C = 3-4+C \implies C = 1.$ Thus $\displaystyle\int \dfrac{x^3+1}{x(x-1)^3}\,dx = -\ln|x| + 2\ln|x-1| -$

$\dfrac{1}{x-1} - \dfrac{1}{(x-1)^2} + C$, where C is now the usual arbitrary constant of integration.

16.   Let $u = x+2$. Then $du = dx$, and $x-1 = u-3$. $\displaystyle\int \dfrac{x-1}{(x+2)^5}\,dx = \int \dfrac{u-3}{u^5}\,du =$

$\displaystyle\int(u^{-4} - 3u^{-5})\,du = \dfrac{u^{-3}}{-3} - \dfrac{3u^{-4}}{-4} + C = -\dfrac{1}{3(x+2)^3} + \dfrac{3}{4(x+2)^4} + C.$

19.   Let $u^3 = x+8$. Then $3u^2 = dx$, $x = u^3-8$. $I = \displaystyle\int \dfrac{\sqrt[3]{x+8}}{x}\,dx = \int \dfrac{u(3u^2)}{u^3-8}\,du =$

$3\displaystyle\int(1 + \dfrac{8}{u^3-8})\,du.$ Now, $u^3-8 = (u-2)(u^2+2u+4) = (u-2)[(u+1)^2+3] =$

$(z-3)(z^2+3)$ where $z = u+1.$

$\dfrac{1}{u^3-8} = \dfrac{1}{(z-3)(z^2+3)} = \dfrac{A}{z-3} + \dfrac{Bz+C}{z^2+3}$. $1 = A(z^2+3) + (Bz+C)(z-3)$.

Equating coefficients gives the system

z^2: $\quad 0 = A + B$
z^1: $\quad 0 = -3B + C$ $\Biggr\}$ $\implies$ $\begin{cases} A = 1/4 \\ B = -1/4 \\ C = -3/4. \end{cases}$
z^0: $\quad 1 = A - C$

Thus $\displaystyle\int \dfrac{1}{u^3-8}\,du = \dfrac{1}{4}\ln|z-3| - \dfrac{1}{8}\ln(z^2+3) - \dfrac{3}{4}\cdot\dfrac{1}{\sqrt{3}}\tan^{-1}\dfrac{z}{\sqrt{3}}$. Replacing z by

u+1, and combining this with the first calculation for I finally gives

$I = 3u + 2\ln|u-2| - \ln(u^2+2u+4) - 2\sqrt{3}\tan^{-1}\dfrac{u+1}{\sqrt{3}} + C$ where $u = \sqrt[3]{x+8}$.

22. $u = \cos(\ln x)$, $dv = dx \implies du = \dfrac{-\sin(\ln x)}{x}\,dx$, $v = x$. $\displaystyle\int \cos(\ln x)\,dx =$

$x\cos(\ln x) + \displaystyle\int \sin(\ln x)\,dx$. Now, $u = \sin(\ln x)$, $dv = dx \implies du = \dfrac{\cos(\ln x)}{x}$,

$v = x$. $\displaystyle\int \cos(\ln x)\,dx = x\cos(\ln x) + x\sin(\ln x) - \displaystyle\int \cos(\ln x)\,dx$.

$\therefore 2\displaystyle\int \cos(\ln x)\,dx = x(\cos(\ln x) + \sin(\ln x)) + C.$

25. With $u = 4-x^2$, $du = -2x\,dx$, $\displaystyle\int \dfrac{x}{\sqrt{4-x^2}}\,dx = -\dfrac{1}{2}\displaystyle\int u^{-1/2}\,du = -u^{1/2} + C = -\sqrt{4-x^2}+C.$

28. $\dfrac{x^3}{x^3-3x^2+9x-27} = 1 + \dfrac{3x^2 - 9x + 27}{(x-3)(x^2+9)}$ and $\dfrac{3x^2 - 9x + 27}{(x-3)(x^2+9)} = \dfrac{A}{x-3} + \dfrac{Bx+C}{x^2+9}$.

$3x^2-9x+27 = A(x^2+9) + (Bx+C)(x-3)$. Equating coefficients yields

x^2: $\quad 3 = A + B$
x^1: $-9 = \quad\quad - 3B + C$ $\Biggr\}$ $\implies$ $\begin{cases} A = 3/2 \\ B = 3/2 \\ C = -9/2. \end{cases}$
x^0: $\quad 27 = 9A \quad\quad - 3C$

Putting this all together at last we get

$\displaystyle\int \dfrac{x^3}{x^3-3x^2+9x-27}\,dx = x + \dfrac{3}{2}\ln|x-3| + \dfrac{3}{4}\ln(x^2+9) - \dfrac{3}{2}\tan^{-1}\dfrac{x}{3} + C.$

31. With $u = e^x$, $du = e^x dx$ and $\displaystyle\int e^x \sec e^x dx = \displaystyle\int \sec u\,du = \ln|\sec u + \tan u|$

$+ C = \ln|\sec e^x + \tan e^x| + C.$

34. Recalling that $\sin 2x = 2\sin x\cos x$, $\displaystyle\int \sin 2x \cos x\,dx = 2\displaystyle\int \sin x \cos^2 x\,dx$

$= -\dfrac{2}{3}\cos^3 x + C.$

37. With $u = e^x$, $du = e^x dx$, $\displaystyle\int e^x\sqrt{1+e^x}\,dx = \displaystyle\int (1+u)^{1/2}\,du = \dfrac{2}{3}(1+u)^{3/2} + C$

$= \dfrac{2}{3}(1+e^x)^{3/2} + C.$

40. $x^2 + 8x + 25 = (x+4)^2 + 9$. So, with $u = x+4$, $du = dx$, $3x+2 = 3(u-4) + 2 =$

$3u = 10$, and $\int \frac{3x+2}{x^2+8x+25}\, dx = \int \frac{3u-10}{u^2+9}\, du = \frac{3}{2}\ln(u^2+9) - \frac{10}{3}\tan^{-1}\frac{u}{3} + C$

$= \frac{3}{2}\ln(x^2+8x+25) - \frac{10}{3}\tan^{-1}\frac{x+4}{3} + C.$

43. $u = x$, $dv = \cot x \csc x\, dx \Longrightarrow du = dx$, $v = -\csc x$. $\int x \cot x \csc x\, dx =$

$-x \csc x + \int \csc x\, dx$.   Now use Formula 15.

46. $u = (\ln x)^2$, $dv = x\, dx \Longrightarrow du = (2 \ln x)/x\, dx$, $v = x^2/2$. $I = \int x(\ln x)^2 dx =$

$(x^2/2)(\ln x)^2 - \int x \ln x\, dx$.   In this integral we use $u = \ln x$, $dv = x\, dx \Longrightarrow$

$du = (1/x)dx$, $v = x^2/2$ to obtain $I = (x^2/2)(\ln x)^2 - [(x^2/2)\ln x - \int x/2\, dx]$

$= (x^2/2)[(\ln x)^2 - \ln x + 1/2] + C.$

49. First write $\int \frac{e^{3x}}{1+e^x}\, dx = \int \frac{e^{2x}}{1+e^x} e^x dx$.   Then let $u = 1+e^x$ so that $du = e^x dx$ and

$e^x = (u-1)$, and we obtain $\int \frac{(u-1)^2}{u}\, du = \int \frac{u^2-2u+1}{u}\, du = \int (u-2+1/u)du =$

$u^2/2 - 2u + \ln|u| + C$, where $u = 1+e^x$.

52. Let $u = 1 + \sin x$.   Then $du = \cos x\, dx$ and $\sin x = u-1$. $\int \frac{\cos^3 x}{\sqrt{1+\sin x}}\, dx$

$= \int \frac{\cos^2 x}{\sqrt{1+\sin x}} \cos x\, dx = \int \frac{1-\sin^2 x}{\sqrt{1+\sin x}} \cos x\, dx = \int \frac{1-(u-1)^2}{\sqrt{u}}\, du$

$= \int \frac{2u-u^2}{u^{1/2}}\, du = \int (2u^{1/2} - u^{3/2})du = \frac{4}{3}u^{3/2} - \frac{2}{5}u^{5/2} + C = \frac{4}{3}(1 + \sin x)^{3/2} -$

$\frac{2}{5}(1 + \sin x)^{5/2} + C.$

55. $\frac{1-2x}{x^2+12x+35} = \frac{1-2x}{(x+5)(x+7)} = \frac{A}{x+5} + \frac{B}{x+7} \Longrightarrow 1-2x = A(x+7) + B(x+5)$.   Setting

$x = -7$ and -5 we get $15 = -2B$ and $11 = 2A$, whence $A = 11/2$, $B = -15/2$.   The

integral then becomes $\int (\frac{11/2}{x+5} - \frac{15/2}{x+7})dx = \frac{11}{2}\ln|x+5| - \frac{15}{2}\ln|x+7| + C.$

58. First let $u = 3x$, $dx = (1/3)du$.   Then use Formulas 73 (with $n = 4$) and 63 to

obtain: $\int \sin^4 3x\, dx = \frac{1}{3}\int \sin^4 u\, du = \frac{1}{3}[-\frac{1}{4}\sin^3 u \cos u + \frac{3}{4}\int \sin^2 u\, du]$

$= -\frac{1}{12}\sin^3 u \cos u + \frac{1}{4}[\frac{1}{2}u - \frac{1}{4}\sin 2u] + C = -\frac{1}{12}\sin^3 3x \cos 3x$

$+ \frac{3}{8}x - \frac{1}{16}\sin 6x + C.$

61. $\int \frac{1}{\sqrt{7+5x^2}}\, dx = \frac{1}{\sqrt{5}}\int \frac{\sqrt{5}\, dx}{\sqrt{7+(\sqrt{5}x)^2}} = \frac{1}{\sqrt{5}}\ln|\sqrt{5}\, x + \sqrt{7+5x^2}| + C$ by Formula 25 with

$u = \sqrt{5}\, x$ and $a^2 = 7$.

64. $\int \cot^5 x \csc x\, dx = \int \cot^4 x(\cot x \csc x)dx = \int (\csc^2 x-1)^2(\cot x \csc x)dx =$

$-\int (u^2-1)^2 du$ (with $u = \csc x$, $du = -\cot x \csc x\ dx$) $= -\int (u^4-2u^2+1)du =$

$-\dfrac{u^5}{5} + \dfrac{2}{3}u^3 - u + C = \dfrac{-\csc^5 x}{5} + \dfrac{2}{3}\csc^3 x - \csc x + C.$

67. $\int (x^2 - \text{sech}^2 4x)dx = \dfrac{x^3}{3} - \dfrac{1}{4}\int \text{sech}^2 u\ du = \dfrac{x^3}{3} - \dfrac{1}{4}\tanh u + C$, where $u = 4x$.

70. Let $u = x^3+1$, $du = 3x^2 dx$ and $x^3 = u-1$. $\int x^5\sqrt{x^3+1}\ dx = (1/3)\int x^3\sqrt{x^3+1}(3x^2)dx$

$= (1/3)\int (u-1)\sqrt{u}\ du = \dfrac{1}{3}\int (u^{3/2}-u^{1/2})du = \dfrac{1}{3}(\dfrac{2}{5}u^{5/2} - \dfrac{2}{3}u^{3/2}) + C$

$= \dfrac{2}{15}(x^3+1)^{5/2} - \dfrac{2}{9}(x^3+1)^{3/2} + C.$

73. Since $\tan u \cos u = \dfrac{\sin u}{\cos u}\cos u = \sin u$, $\int \tan 7x \cos 7x\ dx = \int \sin 7x\ dx =$

$-(1/7)\cos 7x + C.$

76. Let $x = 4\sin\theta$, $dx = 4\cos\theta\ d\theta$, $\sqrt{16-x^2} = 4\cos\theta$. Then $I = \int \dfrac{1}{x^4\sqrt{16-x^2}}\ dx$

$= \int \dfrac{4\cos\theta}{4^4\sin^4\theta(4\cos\theta)}\ d\theta = \dfrac{1}{256}\int \dfrac{1}{\sin^4\theta}\ d\theta = \dfrac{1}{256}\int \csc^4\theta\ d\theta.$ Now, by Formula

78, with $n = 4$, $I = \dfrac{1}{256}[-\dfrac{1}{3}\cot\theta\csc^2\theta + \dfrac{2}{3}\int \csc^2\theta\ d\theta]$

$= \dfrac{1}{256}[-\dfrac{1}{3}\cot\theta\csc^2\theta - \dfrac{2}{3}\cot\theta] + C$

$= -\dfrac{1}{768}\cot\theta(\csc^2\theta + 2) + C = -\dfrac{1}{768}\dfrac{\sqrt{16-x^2}}{x}(\dfrac{16}{x^2} + 2) + C$

$= -\dfrac{\sqrt{16-x^2}(8+x^2)}{384\ x^3} + C.$

79. $\int \dfrac{\sqrt{9-4x^2}}{x^2}\ dx = 2\int \dfrac{\sqrt{9-(2x)^2}}{(2x)^2}2\ dx = 2[-\dfrac{\sqrt{9-4x^2}}{2x} - \sin^{-1}\dfrac{2x}{3}] + C$ by Formula 33

with $u = 2x$, $a^2 = 9.$

82. Let $u = x^2+5$, $du = 2x\ dx$, $x\ dx = (1/2)du$. $\int x(x^2+5)^{3/4}dx = (1/2)\int u^{3/4}du =$

$(1/2)(4/7)u^{7/4} + C = (2/7)(x^2+5)^{7/4} + C.$

85. Let $u = 1 + \cos x$, $du = -\sin x\ dx$. $\int \dfrac{\sin x}{\sqrt{1+\cos x}}\ dx = -\int u^{-1/2}du = -2u^{1/2}$

$+ C = -2\sqrt{1+\cos x} + C.$

88. $\int \sin^4 x \cos^3 x\ dx = \int \sin^4 x(1 - \sin^2 x)\cos x\ dx = \int u^4(1-u^2)du = \int (u^4-u^6)du$

$= (1/5)u^5 - (1/7)u^7 + C$, where $u = \sin x.$

91. $x^4+9x^2+20 = (x^2+4)(x^2+5) \implies f(x) = \dfrac{2x^3+4x^2+10x+13}{(x^2+4)(x^2+5)} = \dfrac{Ax+B}{x^2+4} + \dfrac{Cx+D}{x^2+5} \implies$

$2x^3+4x^2+10x+13 = (Ax+B)(x^2+5) + (Cx+D)(x^2+4) = (A+C)x^3 + (B+D)x^2 +$

$(5A+4C)x + (5B+4D).$ Equating the x^3 and x coefficients, we get the system

$A+C = 2$, $5A +4C = 10$ with solution $A = 2$, $C = 0$. Equating the x^2 and constant

coefficients, we get the system $B+D = 4$, $5B+4D = 13$ with solution $B = -3$,

$D = 7$. Thus $\int f(x)\,dx = \int (\frac{2x}{x^2+4} - \frac{3}{x^2+4} + \frac{7}{x^2+5})\,dx = \ln(x^2+4) - \frac{3}{2}\tan^{-1}\frac{x}{2}$

$+ \frac{7}{\sqrt{5}}\tan^{-1}\frac{x}{\sqrt{5}} + C.$

94. $\int \cot^2 x\, \csc x\, dx = \int (\csc^2 x - 1)\csc x\, dx = \int \csc^3 x\, dx - \int \csc x\, dx =$

$(-(1/2)\csc x \cot x + (1/2)\ln|\csc x - \cot x|) - \ln|\csc x - \cot x| + C =$

$-(1/2)(\csc x \cot x + \ln|\csc x - \cot x|) + C.$ (The integral of $\csc^3 x$ is from #2 or from Formula 72.)

97. Let $u = 2x+3$. Then $x = (1/2)(u-3)$, $dx = (1/2)\,du$. $\int \frac{x^2}{\sqrt[3]{2x+3}}\,dx = \int \frac{(1/4)(u-3)^2}{\sqrt[3]{u}} \cdot$

$(1/2)\,du = \frac{1}{8}\int \frac{u^2-6x+9}{u^{1/3}}\,du = \frac{1}{8}\int (u^{5/3} - 6u^{2/3} + 9u^{-1/3})\,du =$

$\frac{1}{8}[\frac{3}{8}u^{8/3} - 6(\frac{3}{5})u^{5/3} + 9(\frac{3}{2})u^{2/3}] + C$ where $u = 2x+3$.

100. Let $u = x+1$. Then $du = dx$, $u+1 = x+2$. $\int (x+2)^2(x+1)^{10}\,dx = \int (u+1)^2 u^{10}\,du =$

$\int (u^{12}+2u^{11}+u^{10})\,du = (1/13)u^{13} + (2/12)u^{12} + (1/11)u^{11} + C$ where $u = x+1$.

101. $V = 2\pi\int_0^\pi x \sin x\, dx.$

103. $y = \ln \sec x \Longrightarrow y' = \frac{D_x \sec x}{\sec x} = \frac{\sec x \tan x}{\sec x} = \tan x.$

$L_0^{\pi/3} = \int_0^{\pi/3} \sqrt{1 + \tan^2 x}\, dx = \int_0^{\pi/3} \sec x\, dx = \ln|\sec x + \tan x|\,]_0^{\pi/3} = \ln(2+\sqrt{3}).$

105. $m = \int_0^1 (x^2-x^3)\,dx$, $M_x = (1/2)\int_0^1 (x^4-x^6)\,dx$, $M_y = \int_0^1 (x^3-x^4)\,dx.$

106. $m = \int_0^{\pi/2} \cos x\, dx = \sin x\,]_0^{\pi/2} = 1.$ $M_x = \frac{1}{2}\int_0^{\pi/2} \cos^2 x\, dx = \frac{1}{4}\int_0^{\pi/2} (1+\cos 2x)\,dx$

$= \frac{1}{4}[x + \frac{\sin 2x}{2}]_0^{\pi/2} = \frac{\pi}{8}.$ $M_y = \int_0^{\pi/2} x \cos x\, dx.$ Let $u = x$, $dv = \cos x\, dx.$

Then $du = dx$, $v = \sin x$ and $M_y = x \sin x\,]_0^{\pi/2} - \int_0^{\pi/2} \cos x\, dx = \frac{\pi}{2} - 1.$

$\bar{x} = \frac{\pi}{2} - 1$, $\bar{y} = \frac{\pi}{8}.$

107. $m = \pi\int_0^4 x\, dx$, $M_{yz} = \pi\int_0^4 x^2\,dx.$

109. NOTE: we use the formula $\int x\, e^{ax}\, dx = \dfrac{e^{ax}}{a^2}(ax - 1)$ several times here. The shell method is employed to solve the problem.

$$m = 2\pi\int_0^1 x\, e^{-3x}\, dx = \frac{2\pi}{9}\, e^{-3x}(-3x - 1)\Big]_0^1 = \frac{2\pi}{9}(1 - 4e^{-3}). \quad M_{yz} = \frac{2\pi}{2}\int_0^1 x(e^{-3x})^2 dx$$

$$= \pi\int_0^1 xe^{-6x}dx = \frac{\pi}{36}\, e^{-6x}(-6x - 1)\Big]_0^1 = \frac{\pi}{36}(1 - 7e^{-6}). \quad \bar{y} = \frac{\pi(1 - 7e^{-6})}{36} \cdot$$

$$\frac{9}{2\pi(1 - 4e^{-3})} = \frac{1 - 7e^{-6}}{8(1 - 4e^{-3})} \approx 0.15.$$

CHAPTER 11

INDETERMINATE FORMS, IMPROPER INTEGRALS,
AND TAYLOR'S FORMULA

EXERCISES 11.1, page 512

1. $\lim\limits_{x\to 0} \dfrac{\sin x}{2x} = \lim\limits_{x\to 0} \dfrac{\cos x}{2} = \dfrac{\cos 0}{2} = \dfrac{1}{2}$.

4. $\lim\limits_{x\to 4} \dfrac{x-4}{(x+4)^{1/3}-2} = \lim\limits_{x\to 4} \dfrac{1}{(x+4)^{-2/3}/3} = \dfrac{3}{8^{-2/3}} = \dfrac{3}{1/4} = 12.$

7. This function is <u>NOT</u> indeterminate as $x \to 1$ since the denominator does not
    approach 0. Thus L'Hôpital's rule cannot be used. However, by the methods
    of Chapter 2, $\lim\limits_{x\to 1} \dfrac{x^2-3x+2}{x^2-2x-1} = \dfrac{1-3+2}{1-2-1} = \dfrac{0}{-2} = 0.$

10. $\lim\limits_{x\to 0} \dfrac{\sin x}{x - \tan x} = \lim\limits_{x\to 0} \dfrac{\cos x}{1 - \sec^2 x} = -\infty$ since $\cos x \to 1$ and $1 - \sec^2 x$ is negative
    and approaches 0.

13. $\lim\limits_{x\to 0} \dfrac{x - \sin x}{x^3} = \lim\limits_{x\to 0} \dfrac{1 - \cos x}{3x^2} = \lim\limits_{x\to 0} \dfrac{\sin x}{6x} = \lim\limits_{x\to 0} \dfrac{\cos x}{6} = \dfrac{\cos 0}{6} = \dfrac{1}{6}$.

16. $\lim\limits_{x\to 0^+} \dfrac{\cos x}{x} = \infty$ since $\cos x \to 1$ and $x > 0$ as $x \to 0^+$. (Note that L'Hôpital's
    rule cannot be used.)

19. $\lim\limits_{x\to\infty} \dfrac{x^2}{\ln x} = \lim\limits_{x\to\infty} \dfrac{2x}{1/x} = \lim\limits_{x\to\infty} 2x^2 = \infty.$

22. $\lim\limits_{x\to 0} \dfrac{2x}{\tan^{-1}x} = \lim\limits_{x\to 0} \dfrac{2}{1/(1+x^2)} = \dfrac{2}{1} = 2.$

25. The limit is ∞ since $x \cos x + e^{-x} \to 0(1) + 1 = 1$, and $x^2 \to 0$ and is positive
    as $x \to 0$. (Again, L'Hôpital's rule cannot be used.)

28. $\lim\limits_{x\to\infty} \dfrac{x^3+x+1}{3x^3+4} = \lim\limits_{x\to\infty} \dfrac{3x^2+1}{9x^2} = \lim\limits_{x\to\infty} \dfrac{6x}{18x} = \dfrac{1}{3}$.

31. If n is an integer, then after n applications of L'Hôpital's rule (possible,
    since each quotient is the ∞/∞ form), we obtain $\lim\limits_{x\to\infty} \dfrac{x^n}{e^x} = \lim\limits_{x\to\infty} \dfrac{n!}{e^x} = 0.$ If n is
    not an integer, let $k = [n] + 1$ so that $n - k < 0$. After k applications of
    the rule, $\lim\limits_{x\to\infty} \dfrac{x^n}{e^x} = \lim\limits_{x\to\infty} n(n-1)\ldots(n-k+1) \dfrac{x^{n-k}}{e^x} = 0$, since $x^{n-k} \to 0$ and $e^x \to \infty$
    as $x \to \infty$.

34. $\lim\limits_{x\to 0} \dfrac{\sin^2 x + 2\cos x - 2}{\cos^2 x - x\sin x - 1} = \lim\limits_{x\to 0} \dfrac{2\sin x \cos x - 2\sin x}{-2\cos x \sin x - x\cos x - \sin x}$

$$= \lim_{x \to 0} \frac{\sin 2x - 2 \sin x}{-\sin 2x - x \cos x - \sin x} = \lim_{x \to 0} \frac{2 \cos 2x - 2 \cos x}{-2 \cos 2x - 2 \cos x + x \sin x}$$

$$= \frac{2-2}{-2-2+0} = \frac{0}{-4} = 0.$$

37. Using the identity $\dfrac{\tan x - \sin x}{\tan x} = 1 - \dfrac{\sin x}{\sin x / \cos x} = 1 - \cos x$, we have

$$\lim_{x \to 0} \frac{\tan x - \sin x}{x^3 \tan x} = \lim_{x \to 0} \frac{1 - \cos x}{x^3} = \lim_{x \to 0} \frac{\sin x}{3x^2} = \lim_{x \to 0} \frac{\cos x}{6x} \text{ which does not exist}$$

since $\cos x \to 1$, $6x \to 0$.

40. $\displaystyle\lim_{x \to 0} \frac{2 - e^x - e^{-x}}{1 - \cos^2 x} = \lim_{x \to 0} \frac{-e^x + e^{-x}}{2 \cos x \sin x} = \lim_{x \to 0} \frac{-e^x - e^{-x}}{2(\cos^2 x - \sin^2 x)} = \frac{-1-1}{2(1-0)} = \frac{-2}{2} = -1.$

43. $\displaystyle\lim_{x \to 0} \frac{x \sin^{-1} x}{x - \sin x} = \lim_{x \to 0} \frac{x/\sqrt{1-x^2} + \sin^{-1} x}{1 - \cos x}$, which is still of indeterminate form

0/0. Separately, we calculate $D_x \dfrac{x}{\sqrt{1-x^2}} = \dfrac{\sqrt{1-x^2} - x(-2x)/2\sqrt{1-x^2}}{(\sqrt{1-x^2})^2} = \dfrac{(1-x^2)+x^2}{(1-x^2)^{3/2}}$

$= \dfrac{1}{(1+x^2)^{3/2}}$. Continuing the limit calculation from above, using L'Hôpital's

rule, $\displaystyle\lim_{x \to 0} \frac{x/\sqrt{1-x^2} + \sin^{-1} x}{1 - \cos x} = \lim_{x \to 0} \frac{1/(1-x^2)^{3/2} + 1/\sqrt{1-x^2}}{\sin x}$, which does not exist

since the numerator approaches 2 and $\sin x \to 0$ as $x \to 0$.

46. $\displaystyle\lim_{x \to 0} \frac{x - \tan^{-1} x}{x \sin x} = \lim_{x \to 0} \frac{1 - 1/(1+x^2)}{x \cos x + \sin x} = \lim_{x \to 0} \frac{2x/(1+x^2)^2}{-x \sin x + 2 \cos x} = \frac{0}{0+2} = 0.$

49. $\displaystyle\lim_{x \to \infty} \frac{2e^{3x} + \ln x}{e^{3x} + x^2} = \lim_{x \to \infty} \frac{6e^{3x} + 1/x}{3e^{3x} + 2x} = \lim_{x \to \infty} \frac{18e^{3x} - 1/x^2}{9e^{3x} + 2} = \lim_{x \to \infty} \frac{54e^{3x} + 2/x^3}{27e^{3x}}$

$$= \lim_{x \to \infty} \left(2 + \frac{2}{27e^{3x} x^3}\right) = 2 + 0 = 2.$$

EXERCISES 11.2, page 516

1. $\displaystyle\lim_{x \to 0^+} x \ln x = \lim_{x \to 0^+} \frac{\ln x}{1/x} = \lim_{x \to 0^+} \frac{1/x}{-(1/x^2)} = \lim_{x \to 0^+} (-x) = 0.$

4. $\displaystyle\lim_{x \to \infty} x(e^{1/x} - 1) = \lim_{x \to \infty} \frac{e^{1/x} - 1}{1/x} = \lim_{x \to \infty} \frac{(-1/x^2)e^{1/x}}{(-1/x^2)} = \lim_{x \to \infty} e^{1/x} = e^0 = 1.$

7. $\displaystyle\lim_{x \to 0^+} \sin x \ln \sin x = \lim_{x \to 0^+} \frac{\ln \sin x}{\csc x} = \lim_{x \to 0^+} \frac{\cos x / \sin x}{-\csc x \cot x} = \lim_{x \to 0^+} \frac{\cot x}{-\csc x \cot x}$

$$= \lim_{x \to 0^+} \frac{1}{-\csc x} = \lim_{x \to 0^+} (-\sin x) = 0.$$

10. $\lim\limits_{x\to\infty} e^{-x} \ln x = \lim\limits_{x\to\infty} \dfrac{\ln x}{e^x} = \lim\limits_{x\to\infty} \dfrac{1/x}{e^x} = 0.$

13. This is a 1^∞ indeterminate form. If $y = (1 + \dfrac{1}{x})^{5x}$, then $\ln y = 5x \ln(1 + 1/x)$

   $= 5\dfrac{\ln(1 + 1/x)}{1/x}$. $\lim\limits_{x\to\infty} \ln y = \lim\limits_{x\to\infty} \dfrac{5(-1/x^2)/(1+1/x)}{(-1/x^2)} = \lim\limits_{x\to\infty} \dfrac{5}{1+1/x} = 5.$ Thus

   $\lim\limits_{x\to\infty} y = e^5.$

16. If $y = x^x$, then $\ln y = x \ln x$ and $\lim\limits_{x\to 0^+} \ln y = \lim\limits_{x\to 0^+} x \ln x = 0$, by #1. Thus,

   $\lim\limits_{x\to 0^+} y = \lim\limits_{x\to 0^+} x^x = e^0 = 1.$

19. $(\tan x)^x$ is not indeterminate as $x \to \pi/2^-$ since $\tan x \to \infty$ and $x \to \pi/2 > 1.$

   Thus $(\tan x)^x \to \infty$ as $x \to \pi/2^-.$

22. If $y = (1+3x)^{\csc x}$, $\ln y = \csc x \ln(1+3x) = \dfrac{\ln(1+3x)}{\sin x}$.

   $\lim\limits_{x\to 0^+} \ln y = \lim\limits_{x\to 0^+} \dfrac{\ln(1+3x)}{\sin x} = \lim\limits_{x\to 0^+} \dfrac{3}{(1+3x)\cos x} = 3.$ Thus $\lim\limits_{x\to 0^+} y = e^3.$

25. $\lim\limits_{x\to 0} (\dfrac{1}{x} - \dfrac{1}{\sin x}) = \lim\limits_{x\to 0} \dfrac{\sin x - x}{x \sin x} = \lim\limits_{x\to 0} \dfrac{\cos x - 1}{x \cos x + \sin x} = \lim\limits_{x\to 0} \dfrac{-\sin x}{-x \sin x + 2 \cos x}$

   $= \dfrac{0}{0 + 2} = 0.$

28. This is an ∞^0 indeterminate form as $x \to \infty$. $y = (1+e^x)^{e^{-x}} \implies \ln y =$

   $e^{-x} \ln(1+e^x) = \dfrac{\ln(1+e^x)}{e^x}$. $\lim\limits_{x\to\infty} \ln y = \lim\limits_{x\to\infty} \dfrac{e^x/(1+e^x)}{e^x} = \lim\limits_{x\to\infty} \dfrac{1}{1+e^x} = 0.$ Thus

   $\lim\limits_{x\to\infty} y = e^0 = 1.$

31. $\lim\limits_{x\to 0} \cot 2x \tan^{-1}x = \lim\limits_{x\to 0} \dfrac{\tan^{-1}x}{\tan 2x} = \lim\limits_{x\to 0} \dfrac{1/(1+x^2)}{2 \sec^2 2x} = \dfrac{1}{2}$.

34. $\lim\limits_{x\to\infty} (\sqrt{x^2+4} - \tan^{-1}x) = \infty$ since $\sqrt{x^2+4} \to \infty$, but $\tan^{-1}x \to \pi/2$ as $x \to \infty.$

37. $\dfrac{x}{x^2+2x-3} - \dfrac{4}{x+3} = \dfrac{x}{(x+3)(x-1)} - \dfrac{4}{x+3} = \dfrac{1}{x+3}(\dfrac{x}{x-1} - 4) = \dfrac{1}{x+3}(\dfrac{x-4x+4}{x-1}) = \dfrac{-3x+4}{(x+3)(x-1)}$.
   As $x \to -3$, $(-3x+4) \to 13$ and the denominator $\to 0$. Thus the limit does not
   exist. (The right-hand limit is $-\infty$, the left-hand limit is ∞.)

40. $\lim\limits_{x\to\pi/2} \sec x \cos 3x = \lim\limits_{x\to\pi/2} \dfrac{\cos 3x}{\cos x} = \lim\limits_{x\to\pi/2} \dfrac{-3 \sin 3x}{-\sin x} = \dfrac{-3(-1)}{-1} = -3.$

EXERCISES 11.3, page 521

1. $\int_1^\infty \frac{1}{x^{4/3}} \, dx = \lim_{t\to\infty} \int_1^t x^{-4/3} \, dx = \lim_{t\to\infty} -3x^{-1/3} \Big]_1^t = \lim_{t\to\infty} (\frac{-3}{\sqrt[3]{t}} + 3) = 0 + 3 = 3.$

4. $\int_0^\infty \frac{x}{1+x^2} \, dx = \lim_{t\to\infty} \int_0^t \frac{x}{1+x^2} \, dx = \lim_{t\to\infty} \frac{\ln(1+x^2)}{2} \Big]_0^t = \lim_{t\to\infty} (\frac{\ln(1+t^2)}{2} - \frac{\ln 1}{2}) =$

   $\lim_{t\to\infty} \frac{\ln(1+t^2)}{2} = \infty.$ Thus the integral diverges.

7. $\int_0^\infty e^{-2x} \, dx = \lim_{t\to\infty} \int_0^t e^{-2x} \, dx = \lim_{t\to\infty} \frac{e^{-2x}}{-2} \Big]_0^t = \lim_{t\to\infty} (-\frac{1}{2})(e^{-2t} - 1) = \frac{1}{2} .$

10. $\int_0^\infty \frac{1}{\sqrt[3]{x+1}} \, dx = \lim_{t\to\infty} \int_0^t (x+1)^{-1/3} \, dx = \lim_{t\to\infty} (\frac{3}{2})(x+1)^{2/3} \Big]_0^t = \lim_{t\to\infty} \frac{3}{2}[(t+1)^{2/3} - 1]$

    $= \infty.$ Thus the integral diverges.

13. Similar to #4 above.

15. Using (11.7) with a = 0, $\int_{-\infty}^\infty xe^{-x^2} \, dx = \int_{-\infty}^0 xe^{-x^2} \, dx + \int_0^\infty xe^{-x^2} \, dx.$ Now,

    $\int_{-\infty}^0 xe^{-x^2} \, dx = \lim_{t\to-\infty} \frac{-e^{-x^2}}{2} \Big]_t^0 = \lim_{t\to\infty} (\frac{-e^0}{2} + \frac{e^{-t^2}}{2}) = -\frac{1}{2} + 0 = -\frac{1}{2} .$ The second

    integral is $\int_0^\infty xe^{-x^2} \, dx = \lim_{t\to\infty} \frac{-e^{-x^2}}{2} \Big]_0^t = \lim_{t\to\infty} (\frac{-e^{-t^2}}{2} + \frac{e^0}{2}) = 0 + \frac{1}{2} = \frac{1}{2} .$ Since

    both integrals converge, the original integral converges, and its value is

    their sum. Thus $\int_{-\infty}^\infty xe^{-x^2} \, dx = -\frac{1}{2} + \frac{1}{2} = 0.$

16. Using (11.7) with a = 0, $\int_{-\infty}^\infty \cos^2 x \, dx = \int_{-\infty}^0 \cos^2 x \, dx + \int_0^\infty \cos^2 x \, dx.$

    $\lim_{t\to\infty} \int_0^t \cos^2 x \, dx = (\frac{1}{2})\lim_{t\to\infty} \int_0^t (1 + \cos 2x) \, dx = \lim_{t\to\infty} [t + \frac{\sin 2t}{2}] = \infty.$ Since at

    least one of the two improper integrals on the right side diverges, the given
    integral diverges.

19. $\int_0^\infty \cos x \, dx = \lim_{t\to\infty} \sin x \Big]_0^t = \lim_{t\to\infty} \sin t$, which does not exist. Thus, the

    integral diverges.

22. Integrating by parts ($u = x$, $dv = e^{-x} dx$), or using Formula 98, $\int xe^{-x} \, dx =$

    $-(x+1)e^{-x} + C.$ Thus $\int_0^\infty xe^{-x} \, dx = \lim_{t\to\infty} [-(x+1)e^{-x}]_0^t = \lim_{t\to\infty} [-(t+1)e^{-t} + 1]$

$= 0 + 1 = 1.$ $(\lim_{t\to\infty} (t+1)e^{-t} = \lim_{t\to\infty} \frac{t+1}{e^t} = 0$ by L'Hôpital's Rule.$)$

25.   (a) $A = \int_1^\infty \frac{1}{x}\, dx = \lim_{t\to\infty} \int_1^t \frac{1}{x}\, dx = \lim_{t\to\infty} \ln t = \infty.$ Thus no value is assignable

to the area.

(b) $V = \pi \int_1^\infty \frac{1}{x^2}\, dx = \pi \lim_{t\to\infty} \int_1^t x^{-2}\, dx = \pi \lim_{t\to\infty} -\frac{1}{x} \Big]_1^t = \pi.$

28.   (a) $A = \int_8^\infty x^{-2/3}\, dx = \lim_{t\to\infty} 3x^{1/3} \Big]_8^t = \infty.$ Thus no value is assignable to the

area.

(b) $V = \pi \int_8^\infty x^{-4/3}\, dx = \pi \lim_{t\to\infty} [-3x^{-1/3}] \Big]_8^t = (-3\pi) \lim_{t\to\infty} (\frac{1}{t^{1/3}} - \frac{1}{8^{1/3}}) = \frac{3\pi}{2}.$

31.   If f is any odd function $(f(-x) = -f(x))$, defined on $(-\infty,\infty)$, then

$\lim_{t\to\infty} \int_{-t}^t f(x)\,dx$ always exists and, in fact, equals 0. To see this, write

$\int_{-t}^t f(x)\,dx = \int_{-t}^0 f(x)\,dx + \int_0^t f(x)\,dx.$ In the 1st integral let $x = -u$, $dx =$

$-du.$ $\int_{-t}^0 f(x)\,dx = -\int_t^0 f(-u)\,du = -\int_0^t f(u)\,du = -\int_0^t f(x)\,dx.$ Thus $\int_{-t}^t f(x)\,dx$

$= 0$, and the limit is 0. Thus, any odd function f for which $\int_{-\infty}^\infty f(x)\,dx$

diverges will serve as answer. Examples: $f(x) = x$, $f(x) = \sin x$, etc.

34.   Let x be the distance between the electrons and k, the constant of proportionality. Then the force function is $f(x) = k/x^2$. Since the electrons

start 1 cm. apart, $W = \int_1^\infty f(x)\,dx = \int_1^\infty k/x^2\,dx = \lim_{t\to\infty} -\frac{k}{x} \Big]_1^t = \lim_{t\to\infty} [-\frac{k}{t} + k]$

$= 0 + k = k.$

37. $L[\cos x] = \int_0^\infty e^{-sx} \cos x\, dx.$ Using formula 99 from the Table of Integrals

with $a = -s$, $b = 1$, $u = x$, $L[\cos x] = \lim_{t\to\infty} \frac{e^{-sx}}{s^2+1} (-s \cos x + \sin x)] \Big|_0^t =$

$\lim_{t\to\infty} [\frac{e^{-st}}{s^2+1} (-s \cos t + \sin t) - \frac{e^0(-s)}{s^2+1}].$ If $s > 0$, then $\lim_{t\to\infty} e^{-st}(-s \cot +$

$\sin t) = 0$ since for all t, $|-s \cos t + \sin t| \le |-s||\cos t| + |\sin t| \le$

$s+1$, independent of t, whereas $\lim_{t\to 0} e^{-st} = 0.$ Thus $L[\cos x] = \frac{s}{s^2+1}$, $s > 0.$

40. $L[\sin ax] = \int_0^\infty e^{-sx} \sin ax \, dx = \lim_{t \to \infty} \dfrac{e^{-sx}}{s^2+a^2} (-s \sin ax - a \cos ax)]\Big|_0^t =$

$\lim_{t \to \infty} [\dfrac{e^{-st}}{s^2+a^2} (-s \sin at - a \cos at) - \dfrac{e^0(-a)}{s^2+a^2}] = \dfrac{a}{s^2+a^2}$ since the first term $\to 0$

as in #37.

EXERCISES 11.4, page 528

1. Since $1/\sqrt[3]{x}$ is continuous on $[0,8]$ and has an infinite discontinuity at 0, we

have by (11.9), $\int_0^8 1/\sqrt[3]{x} \, dx = \lim_{t \to 0^+} \int_t^8 x^{-1/3} \, dx = \lim_{t \to 0^+} \dfrac{3}{2} x^{2/3}]\Big|_t^8 =$

$\lim_{t \to 0^+} \dfrac{3}{2}(8^{2/3} - t^{2/3}) = \dfrac{3}{2}(2^2 - 0) = 6.$

4. Since $1/(x+2)^{5/4}$ is continuous on $(-2,-1]$ and has an infinite discontinuity at

-2, by (11.9), $\int_{-2}^{-1} 1/(x+2)^{5/4} \, dx = \lim_{x \to -2^+} \int_t^{-1} (x+2)^{-5/4} \, dx = \lim_{t \to -2^+} -4(x+2)^{-1/4}]\Big|_t^{-1}$

$= \lim_{t \to -2^+} [-4 + \dfrac{4}{(t+2)^{1/4}}] = \infty.$ Thus the integral diverges.

7. Since the integrand is discontinuous at $x = 4$, by (11.8), $\int_0^4 \dfrac{1}{(4-x)^{3/2}} \, dx =$

$\lim_{t \to 4^-} \int_0^t (4-x)^{-3/2} \, dx = \lim_{t \to 4^-} \dfrac{2}{(4-x)^{1/2}}]\Big|_0^t = 2 \lim_{t \to 4^-} [\dfrac{1}{(4-t)^{1/2}} - \dfrac{1}{2}] = \infty.$ Thus,

the integral diverges.

10. $\int_1^2 \dfrac{x}{x^2-1} \, dx = \lim_{t \to 1^+} \int_t^2 \dfrac{x}{x^2-1} \, dx = \lim_{t \to 1^+} \dfrac{\ln(x^2-1)}{2}]\Big|_t^2 = \lim_{t \to 1^+} \dfrac{\ln 3 - \ln(t^2-1)}{2} = \infty,$

and the integral diverges.

13. $\int_{-2}^0 \dfrac{1}{\sqrt{4-x^2}} \, dx = \lim_{t \to -2^+} \sin^{-1} \dfrac{x}{2}]\Big|_t^0 = \lim_{t \to -2^+} (-\sin^{-1} \dfrac{t}{2}) = -\sin^{-1}(-1) = \pi/2.$

16. $x^2 - x - 2 = (x-2)(x+1)$. Thus the integrand has a discontinuity at $x = 2$ in

$(0,4)$, and by (11.10) we use $\int_0^4 = \int_0^2 + \int_2^4$ and examine each of the integrals.

As in Sec. 10.4, $\dfrac{1}{(x-2)(x+1)} = \dfrac{1/3}{(x-2)} - \dfrac{1/3}{(x+1)}$. Thus $\int_0^2 \dfrac{1}{x^2-x-2} \, dx =$

$\lim_{t \to 2^-} \dfrac{1}{3}[\ln|x-2| - \ln|x+1|]\Big|_0^t = \dfrac{1}{3} \lim_{t \to 2^-} (\ln|t-2| - \ln 2 - \ln|t+1|) = -\infty,$ and the

original integral diverges since at least one of the integrals on the right

diverges.

19. $\int_0^{\pi/2} \tan x \, dx = \lim_{t \to \pi/2^-} (-\ln|\cos x|)]_0^t = \lim_{t \to \pi/2^-} (-\ln|\cos t|)$. Now, as

$t \to \pi/2^-$, $|\cos t| \to 0$ and $\ln|\cos t| \to -\infty$. Thus the above limit is ∞, and the integral diverges.

22. The integrand, $1/x(\ln x)^2$, is continuous on $[1/e, e]$ except at $x = 1$ where

$\ln x = 0$. Thus, by (11.10) we use $\int_{1/e}^e = \int_{1/e}^1 + \int_1^e$. To get the antiderivative, let $u = \ln x$, $du = (1/x)dx$. Then $\int 1/x(\ln x)^2 dx = \int 1/u^2 \, du = -1/u + C$

$= -1/\ln x + C$. Then the 2nd integral above is $\int_1^e \dfrac{1}{x(\ln x)^2} \, dx = \lim_{t \to 1^+} [- \dfrac{1}{\ln x}]_t^e$

$= \lim_{t \to 1^+} [- \dfrac{1}{\ln e} + \dfrac{1}{\ln t}] = \infty$ since $\ln t \to 0$ and $1/\ln t \to \infty$ as $t \to 1^+$. Since at

at least one of the two integrals above diverges, the original integral diverges.

25. The integrand is continuous on $[0,\pi]$ except at $\pi/2$ since $\pi/2 = 1$. Thus

$\int_0^\pi = \int_0^{\pi/2} + \int_{\pi/2}^\pi$. For the antiderivative, let $u = 1 - \sin x$, $du = -\cos x \, dx$.

$\int \cos x/\sqrt{1 - \sin x} \, dx = -\int u^{-1/2} \, du = 2u^{1/2} + C = 2\sqrt{1 - \sin x} + C$. Thus

$\int_0^{\pi/2} \dfrac{\cos x}{\sqrt{1 - \sin x}} \, dx = \lim_{t \to \pi/2^-} [-2\sqrt{1 - \sin x}]_0^t = \lim_{t \to \pi/2^-} (-2\sqrt{1 - \sin t} + 2) =$

$0 + 2 = 2$. Also, $\int_{\pi/2}^\pi \dfrac{\cos x}{\sqrt{1 - \sin x}} \, dx = \lim_{t \to \pi/2^+} [-2\sqrt{1 - \sin x}]_t^\pi =$

$\lim_{t \to \pi/2^-} (-2 + 2\sqrt{1 - \sin t}) = -2 + 0 = -2$. The given integral is the sum of

the values of these two convergent integrals, namely, $2 + (-2) = 0$.

28. (a) $A = \int_0^1 x^{-1/3} \, dx = \lim_{t \to 0^+} \frac{3}{2} x^{2/3}]_t^1 = \lim_{t \to 0} (\frac{3}{2} - \frac{3}{2}t^{2/3}) = \frac{3}{2}$.

(b) $V = \pi \int_0^1 x^{-2/3} \, dx = \pi \lim_{t \to 0^+} 3x^{1/3}]_t^1 = \pi \lim_{t \to 0^+} (3 - 3t^{1/3}) = 3\pi$.

31. If $n \geq 0$, x^n is continuous on $[0,1]$, and the integral exists by (5.31). If

$n < 0$, x^n has a discontinuity at $x = 0$. If $n \neq -1$, $\int_0^1 x^n \, dx = \lim_{t \to 0^+} \dfrac{x^{n+1}}{n+1}]_t^1 =$

$\lim_{t \to 0^+} \dfrac{t^{n+1} - 1}{n+1}$. If $(n+1) > 0$, i.e., $n > -1$, the limit exists and equals

$1/(n+1)$. If $n < -1$, the limit does not exist and the integral diverges. If

$n = -1$, $\int_0^1 x^{-1}dx = \lim\limits_{t \to 0^+} (-\ln t) = \infty$, and it diverges in this case also. Thus

the integral converges if, and only if, $n > -1$.

EXERCISES 11.5, page 538

1. With $f(x) = \sin x$, $a = \pi/2$, $n = 3$, we have

$$f(x) = \sin x \qquad\qquad f(\pi/2) = 1$$
$$f'(x) = \cos x \qquad\qquad f'(\pi/2) = 0$$
$$f''(x) = -\sin x \qquad\qquad f''(\pi/2) = -1$$
$$f'''(x) = -\cos x \qquad\qquad f'''(\pi/2) = 0$$
$$f^{(4)}(x) = \sin x \qquad\qquad f^{(4)}(z) = \sin z.$$

Remembering to divide $f^{(k)}(a)$ by $k!$, we have

$$\sin x = 1 - \frac{1}{2}(x - \frac{\pi}{2})^2 + \frac{\sin z}{4!}(x - \frac{\pi}{2})^4 \text{ for some } z \text{ between } x \text{ and } \pi/2.$$

4. With $f(x) = e^{-x}$, $a = 1$, $n = 3$, we have

$$f(x) = f''(x) = f^{(4)}(x) = e^{-x} \qquad f(1) = f''(1) = e^{-1}$$
$$f^{(4)}(z) = e^{-z}$$
$$f'(x) = f'''(x) = -e^{-x} \qquad f'(1) = f'''(1) = -e^{-1}$$
$$e^{-x} = e^{-1} - e^{-1}(x-1) + \frac{e^{-1}}{2!}(x-1)^2 - \frac{e^{-1}}{3!}(x-1)^3 + \frac{e^{-z}}{4!}(x-1)^4$$

for some z between x and 1.

7. With $f(x) = 1/x$, $a = -2$, $n = 5$, we have

$$f(x) = 1/x \qquad\qquad f(-2) = -1/2$$
$$f'(x) = -1/x^2 \qquad\qquad f'(-2) = -1/4$$
$$f''(x) = 2/x^3 \qquad\qquad f''(-2) = -2/8$$
$$f'''(x) = -6/x^4 \qquad\qquad f'''(-2) = -6/16$$
$$f^{(4)}(x) = 24/x^5 \qquad\qquad f^{(4)}(-2) = -24/32$$
$$f^{(5)}(x) = -120/x^6 \qquad\qquad f^{(5)}(-2) = -120/64$$
$$f^{(6)}(x) = 720/x^7 \qquad\qquad f^{(6)}(z) = 720/z^7$$

From the entries in the 2nd column, note that $f^{(k)}(-2)/k! = -1/2^{k+1}$ for

$k = 0,1,2,\ldots,5$. Thus $\frac{1}{x} = -\frac{1}{2} - \frac{1}{4}(x+2) - \frac{1}{8}(x+2)^2 - \frac{1}{16}(x+2)^3 - \frac{1}{32}(x+2)^4$

$-\frac{1}{64}(x+2)^5 + \frac{1}{z}(x+2)^6$, for some z between -2 and x.

10. We need here, $\sin \pi/6 = 1/2$, $\csc \pi/6 = 2$, $\cot \pi/6 = \sqrt{3}$. So, with $f(x) =$

$\ln \sin x$, $a = \pi/6$, $n = 3$, we obtain

$$f(x) = \ln \sin x \qquad\qquad f(\pi/6) = \ln(1/2)$$
$$f'(x) = \cos x/\sin x = \cot x \qquad\qquad f'(\pi/6) = \sqrt{3}$$

$$f''(x) = -\csc^2 x \qquad\qquad f''(\pi/6) = -2^2 = -4$$

$$f'''(x) = 2\csc^2 x \cot x \qquad f'''(\pi/6) = 2(4)\sqrt{3} = 8\sqrt{3}$$

$$f^{(4)}(x) = -2(\csc^4 x + 2\csc^2 x \cot^2 x). \quad\text{Thus}$$

$$\ln \sin x = \ln(1/2) + \sqrt{3}(x - \pi/6) - \frac{4}{2!}(x - \pi/6)^2 + \frac{8\sqrt{3}}{3!}(x - \pi/6)^3 +$$

$$\frac{f^{(4)}(z)}{4!}(x - \pi/6)^4 \text{ for some } z \text{ between } \pi/6 \text{ and } x.$$

13. In Maclaurin's formula $a = 0$. So with $n = 4$, we get

$$f(x) = \ln(x+1) \qquad\qquad f(0) = \ln 1 = 0$$

$$f'(x) = (x+1)^{-1} \qquad\qquad f'(0) = 1$$

$$f''(x) = -(x+1)^{-2} \qquad\quad f''(0) = -1$$

$$f'''(x) = 2(x+1)^{-3} \qquad\quad f'''(0) = 2$$

$$f^{(4)}(x) = -6(x+1)^{-4} \qquad f^{(4)}(0) = -6$$

$$f^{(5)}(x) = 24(x+1)^{-5} \qquad f^{(5)}(z) = 24(z+1)^{-5}$$

$$\ln(x+1) = x - \frac{1}{2!}x^2 + \frac{2}{3!}x^3 - \frac{6}{4!}x^4 + \frac{24x^5}{5!(z+1)^5} = x - \frac{1}{2}x^2 + \frac{1}{3}x^3 - \frac{1}{4}x^4 +$$

$$\frac{x^5}{5(z+1)^5} \text{ for some } z \text{ between } x \text{ and } 0.$$

16. $f(x) = \tan^{-1}x \qquad\qquad f(0) = 0$

$$f'(x) = (1+x^2)^{-1} \qquad\qquad f'(0) = 1$$

$$f''(x) = -2x(1+x^2)^{-2} \qquad\quad f''(0) = 0$$

$$f'''(x) = (6x^2-2)(1+x^2)^{-3} \qquad f'''(0) = -2$$

$$f^{(4)}(x) = 24(x-x^3)(1+x^2)^{-4}$$

$$\tan^{-1}x = x - \frac{2}{3!}x^3 + \frac{24(z-z^3)}{4!(1+z^2)^4}x^4 = x - \frac{1}{3}x^3 + \frac{(z-z^3)}{(1+z^2)^4}x^4 \text{ for some } z \text{ between}$$

x and 0.

19. $f(x) = (x-1)^{-2} \qquad\qquad f(0) = (-1)^{-2} = 1$

$$f'(x) = -2(x-1)^{-3} \qquad\qquad f'(0) = -2(-1)^{-3} = 2$$

$$f''(x) = 6(x-1)^{-4} = 3!(x-1)^{-4} \qquad f''(0) = 3!(-1)^{-4} = 3!$$

$$f'''(x) = -24(x-1)^{-5} = -(4!)(x-1)^{-5} \quad f'''(0) = -(4!)(-1)^{-5} = 4!$$

$$f^{(4)}(x) = 120(x-1)^{-6} = 5!(x-1)^{-6} \qquad f^{(4)}(0) = 5!(-1)^{-6} = 5!$$

$$f^{(5)}(x) = -720(x-1)^{-7} = -(6!)(x-1)^{-7} \quad f^{(5)}(0) = -6!(-1)^{-7} = 6!$$

$$f^{(6)}(x) = 5040(x-1)^{-8} = 7!(x-1)^{-8} \qquad f^{(6)}(z) = 7!(z-1)^{-8}$$

From the entries in the second column, we see that $f^{(k)}(0)/k! = (k+1)!/k! = k+1$ for $k = 0,1,\ldots,5$. Thus: $(x-1)^{-2} = 1 + 2x + 3x^2 + 4x^3 + 5x^4 + 6x^5 + 7(z-1)^{-8}x^6$ for some z between 0 and x. (The last coefficient is $f^{(6)}(z)/6!$.)

22. $f(x) = e^{-x^2} \implies f'(x) = -2xe^{-x^2} \implies f''(x) = -2x(-2xe^{-x^2})-2e^{-x^2} =$

$(4x^2-2)e^{-x^2} \implies f'''(x) = (4x^2-2)(-2xe^{-x^2}) + 8xe^{-x^2} = (-8x^3+12x)e^{-x^2} \implies$

$f^{(4)}(x) = (-8x^3+12x)(-2xe^{-x^2}) + (-24x^2+12)e^{-x^2} = (16x^4-48x^2+12)e^{-x^2}$. Thus

$f(0) = 1$, $f'(0) = 0$, $f''(0)/2! = -2/2 = -1$, and $f'''(0) = 0$. Thus:

$e^{-x^2} = 1 - x^2 + (1/4!)(16z^4-48z^2+12)e^{-z^2}x^4$ for some z between 0 and x.

25. Using Exercise 1 with $x = 89^0 = 90^0-1^0 = \frac{\pi}{2} - \frac{\pi}{180}$, we have $x - \frac{\pi}{2} = \frac{-\pi}{180} \approx$

-0.0175 and $\sin 89^0 = 1 - \frac{1}{2}(-\frac{\pi}{180})^2 + \frac{\sin z}{4!} (-\frac{\pi}{180})^4$

$\sin 89^0 \approx 1 - \frac{(.0175)^2}{2} \approx 1 - \frac{.0003}{2} = 0.99985$.

$|R_n(x)| \approx \frac{|\sin z|(.0175)^4}{4!} < \frac{(.02)^4}{24} = \frac{16 \times 10^{-8}}{24} < 10^{-8}$. (Compare the answer

obtained above with the value of $\sin 89^{\underline{0}}$ in a 5-place table. They agree

exactly!)

28. Using Exercise 4 with $x = 1.02$, $x-1 = .02$, we have

$e^{-1.02} = e^{-1}(1 - .02 + \frac{(.02)^2}{2} - \frac{(.02)^3}{6}) + \frac{e^{-z}}{24}(.02)^4$

$e^{-1.02} \approx e^{-1}(1 - .02 + .0002 - .000001) = \frac{0.980199}{e} \approx 0.3606$. Since $1 < z <$

1.02, $e^{-z} < e^{-1} < 1$ and $|R_n(x)| = \frac{e^{-z}}{24}(.02)^4 < \frac{16 \times 10^{-8}}{24} < 10^{-8}$.

31. From Exercise 13, $\ln(x+1) = x - \frac{x^2}{2} + \frac{x^3}{3} - \frac{x^4}{4} + \frac{x^5}{5(z+1)^5}$ for some z between 0

and x. Taking $x = 0.25 = 1/4$ and dropping the remainder, we get

$\ln 1.25 \approx \frac{1}{4} - \frac{1/16}{2} + \frac{1/64}{3} - \frac{1/256}{4} = \frac{1}{4} - \frac{1}{32} + \frac{1}{192} - \frac{1}{1024} = 0.2500 - 0.0312 +$

$0.0052 - 0.0010$ so that $\ln 1.25 \approx 0.2230$. The error in this approximation is

the neglected remainder, $R_4(x)$. Since $0 < z < x = 0.25$, we have $z+1 > 1$ and,

thus, $1/(z+1) < 1$. Now, $R_4(0.25) = \frac{(1/4)^5}{5(z+1)^5}$ so that $0 \leq R_4(0.25) < \frac{(1/4)^5}{5(1)^5} =$

$\frac{1}{5(1024)} = \frac{1}{5120} < 2 \times 10^{-4}$. (3-place accuracy at least.)

34. From Exercise 12, $\log_{10} x = 1 + \frac{1}{10 \ln 10}(x-10) - \frac{1}{200 \ln 10}(x-10)^2 +$

$\frac{1}{3z^3 \ln 10}(x-10)^3$ where z is between 10 and x. Taking $x = 10.01$, so that

$x-10 = .01$, and dropping the remainder, we obtain $\log_{10} 10.01 \approx 1 + \frac{1}{10 \ln 10} \cdot$

$(.01) - \frac{1}{200 \ln 10}(.01)^2 \approx 1.00043$. The error is the neglected remainder

$R_3(10.01) = .01^3/3z^3 \ln 10$. Now, $10 < z < 10.01 \implies 1/z < 10 \implies 1/z^3 < 1/10^3$

$= 10^{-3} \implies 0 \leq R_3(10.01) < (.01^3)10^{-3}/3 \ln 10 = 10^{-9}/3 \ln 10$. Since $\ln 10 >$

2.3, the error is $< 10^{-9}/6.9 \approx 1.4 \times 10^{-10}$. (Thus we'd have 9 place accuracy

if we knew $\ln 10$ exactly to 9 places.)

35. Maclaurin's formula for $\cos x$ with $n - 3$ is $\cos x - 1 - \dfrac{x^2}{2} + \dfrac{(\cos z)x^4}{4!}$

    (Compare #15). Thus if we use $\cos x \approx 1 - \dfrac{x^2}{2}$, the error is $\dfrac{|\cos z||x|^4}{24} \leq$

    $\dfrac{1 \cdot (.1)^4}{24} = \dfrac{10^{-4}}{24} < 5 \times 10^{-6}$. Thus the accuracy is at least 5 places.

37. $f(x) = e^x \implies f^{(k)}(x) = e^x$ and $f^{(k)}(0) = 1$ for all k. Using $n = 2$ in Mac-

    laurin's formula, $e^x = 1 + x + x^2/2 + R_2(x)$ where $R_2(x) = \dfrac{e^z}{3!} x^3$. Thus if we

    neglect the remainder, $e^x \approx 1 + x + x^2/2$ with error $R_2(x)$. Now, $|x| \leq 0.1$

    $\implies -0.1 < x < 0.1$, and since z is between 0 and x, $-0.1 < z < 0.1$ also. Thus

    $e^z < e^{0.1} < e^{\ln 2} = 2$ since $\ln 2 \approx 0.693$ and e^x is an increasing function.

    Thus $|R_2(x)| \leq \dfrac{e^z|x|^3}{6} \leq \dfrac{2(0.1)^3}{6} = \dfrac{10^{-3}}{3} < 5 \times 10^{-4}$ (i.e., 3 place accuracy).

40. $f(x) = f''(x) = f^{(4)}(x) = \cosh x \implies f(0) = f''(0) = 1$, $f^{(4)}(z) = \cosh z$.

    $f'(x) = f'''(x) = \sinh x \implies f'(0) = f'''(0) = 0$. So, with $n = 3$ in Maclaurin's

    formula, $\cosh x = 1 + \dfrac{1}{2}x^2 + R_3(x)$ where $R_3(x) = (\cosh z)x^4/4!$ for some z be-

    between 0 and x. Neglecting the remainder yields the approximation formula.

    The error is $R_3(x)$. Since $\cosh z = (e^z + e^{-z})/2$ and z is between 0 and x, one

    exponent in $\cosh z$ is positive and one is negative. The positive exponential

    is < 2 as in #37 above; the negative exponential is < 1. Thus for such z and

    x, $0 < \cosh z < 3/2$, and $0 < |R_3(x)| \leq \dfrac{(\cosh z)|x|^4}{24} \leq \dfrac{3}{48}(0.1)^4 = \dfrac{1}{16} \times 10^{-4} =$

    6.25×10^{-6} (at least 4 place accuracy).

EXERCISES 11.6, page 542

1. To find $\sqrt[3]{2}$, we must solve $f(x) = x^3 - 2 = 0$. Since $f(1) = -1$ and $f(2) = 6$, the
   root lies between 1 and 2. With the initial guess of $x_1 = 1$ and using $x_{n+1} =$

   $x_n - \dfrac{f(x_n)}{f'(x_n)} = x_n - \dfrac{(x_n^3 - 2)}{3x_n^2}$, we obtain the following results as tabulated,

   after rounding off to 4 places.

n	x_n	$f(x_n)$
1	1.0000	-1.0000
2	1.3333	0.3704
3	1.2639	0.0190
4	1.2599	0.0001
5	1.2599	0.0000

   Thus $\sqrt[3]{2} = 1.2599$ to 4 decimal places, or 1.260 to 3 places.

4. Preliminary analysis: By the methods of Chapter 4, $f(x) = x^3 - 3x + 1$ has a local maximum $f(-1) = 3$ and a local minimum $f(1) = -1$, and f is increasing on $[1,\infty)$. Thus the largest root is > 1 and, since $f(2) = 3$, this root is between 1 and 2. Our initial guess will be $x_1 = 2$ (NOT $x_1 = 1$ since $f'(1) = 0$), and the formula to be used is

$$x_{n+1} = x_n - \frac{(x_n^3 - 3x_n + 1)}{(3x_n^2 - 3)} .$$

n	x_n	$f(x_n)$
1	2.0000	3.0000
2	1.6667	0.6296
3	1.5486	0.0680
4	1.5324	0.0012
5	1.5321	0.0000

Thus the solution is $x = 1.532$ to 3 places.

7. Preliminary analysis: A glance at the graphs of e^{-x} and $\cos x$ shows that there are infinitely many values of $x \geq 0$ where $e^{-x} = \cos x$. These are all solutions of $f(x) = \cos x - e^{-x} = 0$. The smallest positive solution lies between 0 and $\pi/2$. Since $\pi/2 \approx 1.57$, the initial guess will be $x_1 = 1.5$ and the formula is

$$x_{n+1} = x_n - \frac{\cos x_n - e^{-x_n}}{-\sin x_n + e^{-x_n}} .$$

n	x_n	$f(x_n)$
1	1.5000	-0.1524
2	1.3032	-0.0072
3	1.2927	-0.0000

Thus the solution is $x = 1.293$ to 3 places.

10. Preliminary analysis: An in #7, there are infinitely many solutions < 0 now. Because $x < 0 \implies 0 < e^x < 1$ and since the first interval to the left of the origin on which $\sin x$ ranges from 1 to 0 is $(-3/2\pi, -\pi) \approx (-4.71, -3.14)$, the largest solution of $f(x) = e^x - \sin x$ is in this interval. So, we'll guess $x_1 = -4$ and use

$$x_{n+1} = x_n - \frac{e^{x_n} - \sin x_n}{e^{x_n} - \cos x_n} .$$

n	x_n	$f(x_n)$
1	-4.0000	-0.7385
2	-2.9010	+0.2933
3	-3.1868	-0.0386
4	-3.1831	-0.0000

Thus, the solution is x = -3.183 to 3 places.

NOTE: Exercises 7 and 10 illustrate the importance of the preliminary analysis in locating the root approximately. As an experiment, the author tried x_1 = 10 in #10 and obtained the solution -9.4249 after 16 iterations! The guess x_1 = -1 produced the solution -12.5664 after 13 iterations. Both of these numbers are solutions, but not the desired largest one.

13. Preliminary analysis: x = -1 is one root of $f(x) = x^4-x-2 = 0$ obtained by checking the four possible rational roots ±1, ±2. Thus, there is a factor of (x+1), and by long division we find that $x^4-x-2 = (x+1)(x^3-x^2+x-2)$. The cubic factor is negative for large negative x and is positive for large positive x and is increasing on $(-\infty,\infty)$. (Its derivative is $3x^2-2x+1 = 3(x-1/3)^2 + 2/3 \geq 2/3 > 0$.) Thus this factor has exactly one real root so that the original function, x^4-x-2, has exactly one real root in addition to x = -1. Since f(1) = -2 and f(2) = 12, this second root lies between 1 and 2. Our initial guess will be x_1 = 1, and the formula is

$$x_{n+1} = x_n - \frac{x_n^4-x_n-2}{4x_n^3-1}.$$

n	x_n	$f(x_n)$
1	1.0000	-2.0000
2	1.6667	4.0493
3	1.4355	0.8110
4	1.3607	0.0669
5	1.3533	0.0006
6	1.3532	0.0000

Thus the roots are -1 and 1.35 to 2 places.

16. $f(x) = 2e^x + x - 1 \Rightarrow f'(x) = 2e^x + 1 > 0 \Rightarrow$ f is increasing on $(-\infty,\infty)$, and, thus, f has either no roots or exactly one root. Since $f(-1) = 2e^{-1} - 2 < 0$ and f(0) = 2-1 = 1, it has its one and only root between -1 and 0. Our initial guess is x_1 = -1 and the formula is

$$x_{n+1} = x_n - \frac{2e^{x_n} + x_n - 1}{2e^{x_n} + 1}.$$

n	x_n	$f(x_n)$
1	-1.0000	-1.2642
2	-0.2717	0.2520
3	-0.3717	0.0073
4	-0.3748	0.0000

Thus the single root is x = -0.37 to 2 places.

NOTE: Obviously the calculations involved in Newton's method can be quite tedious and cumbersome even if an electronic calculator is available. If the reader has access to a modern digital computer, the work can be considerably shortened. The following program is written in the BASIC language and can be used to solve problems by Newton's method. Lines 10 and 20 contain f(x) and f'(x). (The functions below are those of Exercise 10.) The initial guess is denoted by Z, which is input from the terminal. (The value of Z must be given if batch-processing mode is being used.) The computations stop if f'(x) is zero (line 70), if two consecutive x_n's differ by an amount less than 10^{-5} (line 90), or after 20 iterations (line 105). The form of the output is the same as used in the above solutions. In the program, x_n is Z, x_{n+1} is Y, and the Newton formula is line 80.

```
10   DEF FNA(X) = EXP(X)-SIN(X)
20   DEF FNP(X) = EXP(X)-COS(X)
30   INPUT Z
40   N=1
50   A=FNA(Z), P=FNP(Z)
60   PRINT N,Z,A
70   IF P=0 GOTO 150
80   Y=Z-A/P
90   IF ABS(Y-Z)<.00001 GOTO 130
100  N=N+1
105  IF N>20 GOTO 160
110  Z=Y
120  GOTO 50
130  PRINT N,Y,FNA(Y)
140  END
150  PRINT "F'(X)=ZERO!"
160  END
```

(Some computers may require use of LET in lines 40, 50, 80, 100, and 110; e.g. 40 LET N=1.)

EXERCISES 11.7 (Review), page 543

1. $\lim\limits_{x\to 0} \dfrac{\ln(2-x)}{1 + e^{2x}} = \dfrac{\ln 2}{2}$ by the quotient limit theorem, NOT L'Hôpital's rule.

4. $\lim\limits_{x\to 0} \dfrac{\tan^{-1}x}{\sin^{-1}x} = \lim\limits_{x\to 0} \dfrac{1/(1+x^2)}{1/\sqrt{1-x^2}} = \dfrac{1}{1} = 1.$

7. $\lim\limits_{x\to\infty} \dfrac{x^e}{e^x} = \lim\limits_{x\to\infty} \dfrac{ex^{e-1}}{e^x} = \lim\limits_{x\to\infty} \dfrac{e(e-1)x^{e-2}}{e^x} = \lim\limits_{x\to\infty} \dfrac{e(e-1)(e-2)x^{e-3}}{e^x} = 0$ since $x^{e-3} \to 0$

   and $e^x \to \infty$ as $x \to \infty$. (See also #31, Sec. 11.1.)

10. $\lim\limits_{x\to 0} \tan^{-1}x \csc x = \lim\limits_{x\to 0} \dfrac{\tan^{-1}x}{\sin x} = \lim\limits_{x\to 0} \dfrac{1/(1+x^2)}{\cos x} = \dfrac{1}{1} = 1.$

13. $y = (e^x + 1)^{1/x} \implies \ln y = \dfrac{\ln(e^x+1)}{x}$. $\lim\limits_{x\to\infty} \ln y = \lim\limits_{x\to\infty} \dfrac{e^x}{e^x+1} = \lim\limits_{x\to\infty} \dfrac{e^x}{e^x} = 1.$

   Thus $\lim\limits_{x\to\infty} y = \lim\limits_{x\to\infty} e^{\ln y} = e^1 = e.$

16. $\displaystyle\int_4^\infty \dfrac{1}{x\sqrt{x}}\,dx = \lim\limits_{t\to\infty} \left[-2x^{-1/2}\right]_4^t = -2 \lim\limits_{t\to\infty} \left[\dfrac{1}{\sqrt{t}} - \dfrac{1}{2}\right] = 1.$

19. With the discontinuity of the integrand at $x = 0$, we write $\displaystyle\int_{-8}^1 = \int_{-8}^0 + \int_0^1$.

   The 1st of these is $\lim\limits_{t\to 0^-} \displaystyle\int_{-8}^t x^{-1/3}\,dx = \lim\limits_{t\to 0^-} \dfrac{3}{2}x^{2/3}\Big]_{-8}^t = \dfrac{3}{2}[0-(-8)^{2/3}] = -6.$

   The second is $\lim\limits_{t\to 0^+} \displaystyle\int_t^1 x^{-1/3}\,dx = \dfrac{3}{2}$. Thus the integral converges and has

   value $-6 + \dfrac{3}{2} = -\dfrac{9}{2}$.

22. $\displaystyle\int_1^2 \dfrac{1}{x\sqrt{x^2-1}}\,dx = \lim\limits_{t\to 1^+} \sec^{-1}x\Big]_t^2 = \sec^{-1}2 - \sec^{-1}1 = \dfrac{\pi}{3} - 0 = \dfrac{\pi}{3}$.

25. $\displaystyle\int_0^1 \dfrac{\ln x}{x}\,dx = \lim\limits_{t\to 0^+} \dfrac{1}{2}\ln^2 x\Big]_t^1 = \lim\limits_{t\to 0^+} \left(-\dfrac{1}{2}\right)\ln^2 t = -\infty.$ Thus the integral diverges.

28. (a) $f(x) = e^{-x^3}$ $\qquad\qquad\qquad\qquad f(0) = 1$

   $f'(x) = -3x^2 e^{-x^3}$ $\qquad\qquad\qquad f'(0) = 0$

   $f''(x) = (9x^4-6x)e^{-x^3}$ $\qquad\qquad f''(0) = 0$

   $f'''(x) = (-27x^6+54x^3-6)e^{-x^3}$ $\qquad f'''(0) = -6$

   $f^{(4)}(x) = (81x^8-324x^5+180x^2)e^{-x^3}$

   $e^{-x^3} = 1 - x^3 + \dfrac{(81z^8-324z^5+180z^2)e^{-z^3}}{24}x^4$ for some z between 0 and x.

(b) If $f(x) = 1/(1-x)$, then $f^{(k)}(x) = k!/(1-x)^{k+1}$ by an easy calculation or by #45, Sec. 3.9. It follows that $f^{(k)}(0) = k!$, and $f^{(k)}(0)/k! = 1$. Thus $\frac{1}{1-x} = 1 + x + x^2 + x^3 + x^4 + x^5 + x^6 + \frac{x^7}{(1-z)^8}$ for some z between 0 and x.

31. Preliminary analysis: Since $f'(x) = 6x^2 - 8x - 3 > 0$ for $x > 2$, and since $f(2) = -5$, $f(3) = 10$, it follows that the largest root of $f(x) = 2x^3 - 4x^2 - 3x + 1 = 0$ is between 2 and 3. Using $x_1 = 2$ and $x_{n+1} = x_n - \frac{2x_n^3 - 4x_n^2 - 3x_n + 1}{6x_n^2 - 8x_n - 3}$ we have:

n	x_n	$f(x_n)$
1	2.0000	-5.0000
2	3.0000	10.0000
3	2.6296	1.8188
4	2.5254	0.1257
5	2.5171	0.0008
6	2.5170	0.0000

Thus, the solution is x = 2.5170 to 4 places.

CHAPTER 12

INFINITE SERIES

1. $a_n = \dfrac{n}{3n+2} \implies a_1 = \dfrac{1}{3+2} = \dfrac{1}{5}$, $a_2 = \dfrac{2}{6+2} = \dfrac{1}{4}$, $a_3 = \dfrac{3}{9+2} = \dfrac{3}{11}$, $a_4 = \dfrac{4}{12+2} = \dfrac{2}{7}$.

$\lim\limits_{n\to\infty} a_n = \lim\limits_{n\to\infty} \dfrac{n}{3n+2} = \lim\limits_{n\to\infty} \dfrac{1}{3 + 2/n} = \dfrac{1}{3+0} = \dfrac{1}{3}$.

4. $a_n = \dfrac{4}{8-7n} \implies a_1 = \dfrac{4}{8-7} = 4$, $a_2 = \dfrac{4}{8-14} = -\dfrac{2}{3}$, $a_3 = \dfrac{4}{8-21} = -\dfrac{4}{13}$, $a_4 = \dfrac{4}{8-28} =$

$-\dfrac{1}{5}$. $\lim\limits_{n\to\infty} \dfrac{4}{8-7n} = \lim\limits_{n\to\infty} \dfrac{4/n}{8/n - 7} = \dfrac{0}{-7} = 0$.

7. $a_1 = \dfrac{1\cdot 4}{2} = 2$, $a_2 = \dfrac{3\cdot 7}{9} = \dfrac{7}{3}$, $a_3 = \dfrac{5\cdot 10}{28} = \dfrac{25}{14}$, $a_4 = \dfrac{7\cdot 13}{65} = \dfrac{7}{5}$.

$\lim\limits_{n\to\infty} \dfrac{(2n-1)(3n+1)}{n^3+1} = \lim\limits_{n\to\infty} \dfrac{(1/n)(2-1/n)(3+1/n)}{1+1/n^3} = \dfrac{0}{1} = 0$.

10. $a_1 = \dfrac{100}{5} = 20$, $a_2 = \dfrac{200}{2\sqrt{2} + 4} = \dfrac{100}{\sqrt{2} + 2}$, $a_3 = \dfrac{300}{3\sqrt{3} + 4}$, $a_4 = \dfrac{400}{8+4} = \dfrac{100}{3}$.

$\lim\limits_{n\to\infty} \dfrac{100n}{n^{3/2} + 4} = \lim\limits_{n\to\infty} \dfrac{100/\sqrt{n}}{1 + 4/n\sqrt{n}} = 0$.

13. $a_1 = 1 + .1 = 1.1$, $a_2 = 1 + (.1)^2 = 1.01$, $a_3 = 1 + (.1)^3 = 1.001$, $a_4 = 1 +$

$(.1)^4 = 1.0001$. $\lim\limits_{n\to\infty} 1 + (.1)^n = 1 + 0 = 1$.

16. $a_1 = 2$, $a_2 = 3/\sqrt{2}$, $a_3 = 4/\sqrt{3}$, $a_4 = 5/2$. $\lim\limits_{n\to\infty} \dfrac{n+1}{\sqrt{n}} = \lim\limits_{n\to\infty} \sqrt{n} + \dfrac{1}{\sqrt{n}} = \infty$.

19. $\lim\limits_{n\to\infty} \arctan n = \pi/2$ by (12.6) and the definition of the function, arctan x.

See also Figure 9.9.

22. $\lim\limits_{n\to\infty} \dfrac{1.0001^n}{1000} = \infty$ by (12.5) with $r = 1.001 > 1$.

25. Let $f(x) = (4x^4+1)/(2x^2+1)$, an ∞/∞ form as $x \to \infty$. $\lim\limits_{x\to\infty} f(x) = \lim\limits_{x\to\infty} \dfrac{16x^3}{4x} = \infty$

by L'Hôpital's rule. Thus $\lim\limits_{n\to\infty} f(x) = \infty$ also by (12.6).

28. Let $f(x) = e^{-x} \ln x = \ln x/e^x$. Then $\lim\limits_{x\to\infty} \dfrac{\ln x}{e^x} = \lim\limits_{x\to\infty} \dfrac{1/x}{e^x} = 0$. Thus $\lim\limits_{n\to\infty} e^{-n} \ln n$

$= 0$ also.

31. Since $|\sin n| \le 1$, $|2^{-n} \sin n| \le 2^{-n} = (1/2)^n$, and $\lim\limits_{n\to\infty} (1/2)^n = 0$, it follows

that $\lim\limits_{n\to\infty} |2^{-n} \sin n| = 0$ by (12.7). Then by (12.8), $\lim\limits_{n\to\infty} 2^{-n} \sin n = 0$.

34. Let $f(x) = x \sin(1/x)$, an $\infty\cdot 0$ indeterminate form as $x \to \infty$. $\lim\limits_{x\to\infty} f(x) =$

$$\lim_{x \to \infty} \frac{\sin(1/x)}{1/x} = \lim_{x \to \infty} \frac{(-1/x^2)\cos(1/x)}{(-1/x^2)} = \lim_{x \to \infty} \cos \frac{1}{x} = \cos 0 = 1.$$

Thus $\lim_{n \to \infty} n \sin(1/n) = 1$.

37. $k \geq 2 \implies 0 < \frac{1}{k^n} \leq \frac{1}{2^n}$, and $\frac{1}{2^n} \to 0$ by (12.5i).

40. Hint: b is an upper bound for (a,b). By (12.10), there is a least upper bound, say, v, and $v \leq b$. If $v < b$, show that you can find a number w between v and b. Then w is in (a,b) but larger than the upper bound v, a contradiction. Thus v = b.

EXERCISES 12.2, page 561

1. Since a = 3, r = 1/4 < 1, it converges with sum $3/(1 - \frac{1}{4}) = 4$.

4. Since a = 1, r = e/3 < 1, it converges with sum 1/(1-e/3) = 3/(3-e).

7. Writing $\sum_{n=1}^{\infty} 2^{-n}3^{n-1} = \frac{1}{2} + \frac{3}{2^2} + \frac{3^2}{2^3} + \ldots = \frac{1}{2}(1 + \frac{3}{2} + (\frac{3}{2})^2 + \ldots)$ it diverges since r = 3/2 > 1.

10. The series diverges since $r = \sqrt{2} > 1$.

13. The terms are obtained by multiplying those in Example 1 by 5. Thus it converges by (12.17(ii)) with c = 5.

16. Since $\lim_{n \to \infty} \frac{n}{n+1} = 1 \neq 0$, the series diverges by (12.14).

19. Since $\lim_{n \to \infty} \frac{1}{\sqrt[n]{e}} = \lim_{n \to \infty} e^{-1/n} = e^0 = 1 \neq 0$, the series diverges by (12.14).

22. Since $\sum \frac{1}{3^n}$ and $\sum \frac{(-1)}{4^n}$ both converge, as geometric series with r = 1/3, 1/4, respectively, the given series converges by (12.17(ii)).

25. $\sum (3/2)^n$ is a divergent geometric series (r = 3/2), and $\sum (2/3)^n$ is a convergent geometric series (r = 2/3). Thus the given series diverges by (12.18).

28. $\sum 2^{-n} = \sum (1/2)^n$ and $\sum 2^{-3n} = \sum (2^{-3})^n = \sum (1/8)^n$ both converge as geometric series with r = 1/2 and 1/8, respectively. Thus the given series converges by (12.17(iii)).

31. If $\sum ca_n$ converged, so would $\sum(1/c)ca_n = \sum a_n$ by (12.17(ii)). This contradicts the assumed divergence of $\sum a_n$.

34. The given series, $\sum(-1)^{n+1}$ has terms $(-1)^{n+1}$ and partial sums $S_1 = 1$, $S_2 = 0$, $S_3 = 1$, $S_4 = 0$, etc., and the series diverges both by (12.14) and the fact that $\lim_{n\to\infty} S_n$ does not exist. When grouping is done as shown in the problem, an entirely new series results. Its terms are $b_1 = a_1 + a_2$, $b_2 = a_3 + a_4$, etc. so that every term is 0 and, thus, every partial sum of this different series is 0. Thus, it is not the original series that has sum 0, but rather, it is the new series, obtained by regrouping, that has sum 0.

37. $3.2\overline{394} = 3.2 + \dfrac{394}{10^4} + \dfrac{394}{10^7} + \dfrac{394}{10^{10}} + \ldots = 3.2 + \dfrac{1}{10^4}(394 + \dfrac{394}{10^3} + \dfrac{394}{10^6} + \ldots)$

$= 3.2 + \dfrac{1}{10^4} \cdot \dfrac{394}{1-10^{-3}}$ by (12.12) with $a = 394$ and $r = 1/10^3 = 10^{-3}$. Thus

$3.2\overline{394} = 3.2 + \dfrac{1}{10^4} \cdot \dfrac{394}{999/1000} = 3.2 + \dfrac{394}{9,990} = \dfrac{3.2(9990)+394}{9990} = \dfrac{32,362}{9,990}$.

40. The total distance is approximately $24 + 24(\frac{5}{6}) + 24(\frac{5}{6})^2 + \ldots = \dfrac{24}{1-5/6} = \dfrac{24}{1/6} = 144$ cm.

EXERCISES 12.3, page 571

1. $f(x) = (3+2x)^{-2}$ is > 0, continuous, and decreasing on $[1,\infty)$. (Decreasing since $f'(x) = -4(3+2x)^{-2} < 0$ there.) Since $\displaystyle\int_1^\infty (3+2x)^{-2}\, dx = \lim_{t\to\infty} (-\frac{1}{2})\frac{1}{3+2x}\Big]_1^t$ $= \dfrac{1}{10}$, the given series converges by the integral test (12.20).

4. The integral test can apply to a series $\displaystyle\sum_{n=2}^{\infty} f(n)$ if $f(x)$ satisfies the hypotheses of (12.20) on $[2,\infty)$. The same proof works after a change of variable $u = \ln x$. Here, $f(x) = 1/x(\ln x)^2$ satisfies all 3 conditions on $[2,\infty)$. (f is decreasing either by computing $f'(x)$ or by the fact that x and $\ln x$ are increasing.) Since $\displaystyle\int_2^\infty \dfrac{1}{x(\ln x)^2}\, dx = \lim_{t\to\infty} -\dfrac{1}{\ln x}\Big]_2^t = \dfrac{1}{\ln 2}$, the given series converges.

7. $f(x) = (2x+1)^{-1/3}$ is > 0, continuous and decreasing on $[1,\infty)$. $(f'(x) = -(2/3)(2x+1)^{-4/3} < 0$ there.) Since $\displaystyle\int_1^\infty (2x+1)^{-1/3}\, dx = \lim_{t\to\infty} \frac{3}{4}[(2t+1)^{2/3} - 3^{2/3}]$ $= \infty$, the series diverges.

10. $f(x) = xe^{-x}$ is > 0, continuous and decreasing on $[1,\infty)$. $(f'(x) = (1-x)e^{-x} < 0$ if $x > 1$.) Since $\displaystyle\int_1^\infty xe^{-x}\, dx = \lim_{t\to\infty} (-te^{-t} - e^{-t} + 2e^{-1}) = 2e^{-1}$, the

series converges.

13. $f(x) = x2^{-x^2}$ satisfies the hypotheses of (2.20). (It's decreasing on $[1,\infty)$

since $f'(x) = (1-2x^2)(\ln 2)2^{-x^2} < 0$ if $x \geq 1$.) Since $\int_1^\infty x2^{-x^2} dx =$

$\lim_{t\to\infty} \frac{1}{2 \ln 2} (-2^{-t^2}+2^{-1}) = \frac{1}{4 \ln 2}$, the series converges.

16. For large n, $n\sqrt{n^2-1} \approx n\sqrt{n^2} = n^2$. Thus, with $a_n = 1/n\sqrt{n^2-1}$ we select $b_n = 1/n^2$

to obtain $\lim_{n\to\infty} \frac{a_n}{b_n} = \lim_{n\to\infty} \frac{n^2}{n\sqrt{n^2-1}} = \lim_{n\to\infty} \frac{1}{\sqrt{1-1/n^2}} = 1$. Since $\sum b_n = \sum 1/n^2$ is a

p-series with p = 2, $\sum b_n$ converges by (12.21(ii)) and, by the Limit Comparison

Test (12.23), so does the given series, $\sum a_n$.

19. Since $\frac{1}{n3^n} \leq \frac{1}{3^n}$ for $n \geq 1$, and since $\sum (\frac{1}{3})^n$ converges as a geometric series

with r = 1/3, the series $\sum \frac{1}{n3^n}$ converges by the Comparison Test (12.22).

22. For large n, $\sqrt{n}$ is much larger than 3, and it appears that $2/(3+\sqrt{n})$ behaves

very much like $2/\sqrt{n}$. This suggests that we compare the terms of the given

series with $1/\sqrt{n}$ (or $2/\sqrt{n}$). Thus with $a_n = 2/(3+\sqrt{n})$ and $b_n = 1/\sqrt{n}$,

$\lim_{n\to\infty} \frac{a_n}{b_n} = \lim_{n\to\infty} \frac{2\sqrt{n}}{3+\sqrt{n}} = 2$. But $\sum b_n = \sum 1/\sqrt{n}$ diverges since it is a p-series with

p = 1/2. Thus the given series diverges by the Limit Comparison Test.

25. For large n, neglecting the smaller powers of n, $\sqrt{4n^3-5n} \approx \sqrt{4n^3} = 2n^{3/2}$. So,

with $a_n = 1/\sqrt{4n^3-5n}$, we select $b_n = 1/2n^{3/2}$ and obtain $\lim_{n\to\infty} \frac{a_n}{b_n} =$

$\lim_{n\to\infty} \frac{2n^{3/2}}{\sqrt{4n^3-5n}} = \lim_{n\to\infty} \frac{2}{\sqrt{4-5/n^2}} = \frac{2}{\sqrt{4}} = 1$ (having divided numerator and denominator

by $n^{3/2} = \sqrt{n^3}$). Since $\sum b_n$ is $\frac{1}{2}$ times a p-series with p = 3/2, $\sum b_n$ converges,

and, by the Limit Comparison Test, so does $\sum a_n$.

28. $0 \leq \sin^2 n \leq 1 \implies 0 \leq \sin^2 n/2^n \leq 1/2^n$. Since $\sum 1/2^n$ converges (geometric

series, r = 1/2), the given series converges by the Comparison Test.

31. $-1 \leq \cos n \leq 1 \implies 1 \leq 2 + \cos n \leq 3 \implies \frac{2 + \cos n}{n^2} \leq \frac{3}{n^2}$. Since $\sum 3/n^2$ con-

verges as a p-series with p = 2, the given series converges by the Comparison

Test.

34. As above, if we retain the largest power of n inside the radical, it appears

that the terms of the series behave like $1/\sqrt{n^3} = 1/n^{3/2}$. Thus with

$a_n = 1/\sqrt{n(n+1)(n+2)}$ and $b_n = 1/n^{3/2}$, $\lim\limits_{n\to\infty} a_n/b_n = 1$. But $\sum b_n$ converges as a

p-series with $p = 3/2$. Thus the given series converges by the Limit

Comparison Test.

37. $n \geq 2 \implies 1/n^n \leq 1/n^2$. Since $\sum 1/n^2$ converges ($p = 2$), $\sum\limits_{n=2}^{\infty} 1/n^n$ converges

by the Comparison Test. Since this differs from the given series only by

deleting the 1st term, the given series converges by (12.16).

40. Since $n \geq 1 \implies \ln n \geq 0$, we have $a_n = \dfrac{n + \ln n}{n^2+1} \geq \dfrac{n}{n^2+1}$. The series

$\sum n/(n^2+1)$ diverges either by the integral test or by the Limit Comparison

Test using $b_n = 1/n$. Thus by the Comparison Test, 2nd conclusion, the given

series diverges.

43. For sufficiently large n, $\ln n < n$ since $\lim\limits_{n\to\infty} \dfrac{\ln n}{n} = 0$ (by L'Hôpital's Rule).

(In fact, it can be shown that $\ln n < n$ for all $n \geq 1$.) Thus $\dfrac{\ln n}{n^3} < \dfrac{n}{n^3} = \dfrac{1}{n^2}$.

Since $\sum 1/n^2$ converges ($p = 2$) so does the given series by the Comparison

Test.

46. For $n \geq 1$, $-1 \leq \sin n \leq 1$ and $2 \leq 2^n$. Thus $1 \leq \sin n + 2^n \leq 1 + 2^n$ and

$\dfrac{\sin n + 2^n}{n + 5^n} \leq \dfrac{1 + 2^n}{5^n} = (\tfrac{1}{5})^n + (\tfrac{2}{5})^n$. Since both $\sum(1/5)^n$ and $\sum(2/5)^n$ converge

(geometric series, $r = 1/5$, $2/5$), the given series converges by the Comparison

Test.

49. Hint: For large n, $a_n/b_n < 1$, or $a_n < b_n$.

50. Hint: For some N, $n \geq N \implies a_n/b_n > 1$, or $a_n > b_n$. If S_n and T_n are the nth

partial sums of $\sum a_n$ and $\sum b_n$, respectively, and if $n > N$, $S_n = (S_n - S_N) +$

$S_N > (T_n - S_N) + S_N$. What does the divergence of $\sum b_n$ tell you about T_n as

$n \to \infty$?

EXERCISES 12.4, page 575

1. $2(n+1)+1 > 2n+1 \implies \dfrac{1}{\sqrt{2(n+1)+1}} < \dfrac{1}{\sqrt{2n+1}} \implies a_{n+1} < a_n$. Since $\lim\limits_{n\to\infty} a_n =$

$\lim\limits_{n\to\infty} 1/\sqrt{2n+1} = 0$, the series converges by the Alternating Series Test (AST).

4. With $f(x) = \dfrac{x}{x^2+4}$, $f'(x) = \dfrac{4-x^2}{(x^2+4)^2} < 0$ for $x > 2$, and the terms of the series

are decreasing for $n \geq 2$. Since $n/(n^2+4) \to 0$ as $n \to \infty$, the given series con-

verges by the AST.

7. With $f(x) = \frac{x^2+1}{x^3+1}$, $f'(x) = \frac{-x(x^3+3x-2)}{(x^3+1)^2}$. The factor (x^3+3x-2) has value 2

    when $x = 1$ and increases thereafter. (Its derivative is $3x^2+3 \geq 3 > 0$.)

    Thus the numerator of $f'(x)$ is < 0 for $x \geq 1$ and $f'(x) < 0$. Thus the terms

    of the series are decreasing for $n \geq 1$ and $\lim\limits_{n\to\infty} \frac{n^2+1}{n^3+1} = 0$. Thus the given

    series converges by the AST.

8. Since $\lim\limits_{n\to\infty} a_n = \lim\limits_{n\to\infty} \frac{3n+4}{5n+7} = \frac{3}{5} \neq 0$, the series diverges.

10. With $f(x) = \frac{\sqrt{x+1}}{8x+5}$, $f'(x) = \frac{-8x-11}{2\sqrt{x+1}(8x+5)^2} < 0$ for $x \geq 1$. Thus the terms of the

    series decrease, and $\lim\limits_{n\to\infty} a_n = 0$.   Thus convergence by the AST.

13. We first note that the series converges by the AST since $a_n = 1/n! \to 0$ and a_{n+1}

    $= 1/(n+1)! = 1/(n+1)n! < 1/n! = a_n$. Recall also that if the sum, S, is approx-

    imated by the nth partial sum, S_n, the error is no more than a_{n+1}. Since we

    want 3 decimal place accuracy, the error is to be less than 5×10^{-4}. Thus we

    seek n such that $a_{n+1} = 1/(n+1)! < 5\times10^{-4}$, or, equivalently $(n+1)! > 10^4/5 =$

    2000. With $n = 6$, $(n+1)! = 7! = 5040$, whereas with $n = 5$, $6! = 720$. Thus, to

    3 place accuracy $S \approx S_6 = 1-1+\frac{1}{2}-\frac{1}{6}+\frac{1}{24}-\frac{1}{120}+\frac{1}{720} \approx 0.500-0.1667+0.0417-0.0083+$

    $0.0014 = 0.3681 \approx 0.368$. (Recall above that $0! = 1$.)

16. Here, we seek n such that $a_{n+1} = 1/(n+1)^5 < 5 \times 10^{-4}$ or $(n+1)^5 > 2000$. With

    $n = 4$, $(n+1)^5 = 5^5 = 3125$. So, to 3 places $S \approx S_4 = 1 - \frac{1}{2^5} + \frac{1}{3^5} - \frac{1}{4^5} =$

    $1 - \frac{1}{32} + \frac{1}{343} - \frac{1}{1024} \approx 1 - 0.0313 + 0.0041 - 0.0010 = 0.9718 \approx 0.972$.

19. With $a_n = 1/n^n$, we seek the smallest n such that $a_{n+1} = 1/(n+1)^{n+1} < 5 \times 10^{-5}$

    for 4 place accuracy. This inequality is equivalent to $(n+1)^{n+1} > 10^5/5 =$

    $20,000$. Since $5^5 = 3125$ and $6^6 = 46,656$, we obtain $n+1 = 6$ or $n = 5$.

EXERCISES 12.5, page 581

1. The given series converges by the AST. However, $\sum|a_n| = \sum|(-1)^{n+1}/\sqrt{n}| =$

    $\sum 1/\sqrt{n}$ which is a divergent p-series $(p = 1/2)$. Thus the given series con-

    verges conditionally.

4. $\sum|a_n| = \sum|(-1)^n e^{-n}| = \sum e^{-n} = \sum(1/e)^n$ which is a convergent geometric series

    $(r = 1/e < 1)$. Thus the given series converges absolutely.

7. With $a_n = (-1)^{n-1} \frac{(3n+1)}{2^n}$, $\lim\limits_{n\to\infty} \left|\frac{a_{n+1}}{a_n}\right| = \lim\limits_{n\to\infty} \frac{3n+4}{2^{n+1}} \cdot \frac{2^n}{3n+1} = \lim\limits_{n\to\infty} \frac{1}{2} \frac{3n+4}{3n+1} = \frac{1}{2}$.

Since this limit is < 1, the given series converges absolutely by the ratio test.

10. $\lim\limits_{n\to\infty} \left|\frac{a_{n+1}}{a_n}\right| = \lim\limits_{n\to\infty} \frac{2^n}{5^{n+1}(n+2)} \cdot \frac{5^n(n+1)}{2^{n-1}} = \lim\limits_{n\to\infty} \frac{2}{5} \left(\frac{n+1}{n+2}\right) = \frac{2}{5} < 1$. Thus absolute convergence by the ratio test.

13. Since $|a_n| = \frac{\sqrt{n}}{n^2+1} < \frac{\sqrt{n}}{n^2} = \frac{1}{n^{3/2}}$, and since $\sum \frac{1}{n^{3/2}}$ converges (p = 3/2), the given series converges absolutely.

14. The given series converges by the AST. However, $|a_n| = \frac{\sqrt{n}}{n+1}$, and $\sum |a_n|$ diverges by the limit comparison test choosing $b_n = 1/\sqrt{n}$ (p = 1/2). Thus the given series converges conditionally.

16. $|a_n| = (2n+1)/(n^2+n^3)$, and $\sum |a_n|$ converges by the limit comparison test choosing $b_n = 1/n^2$ (p = 2). Thus it converges absolutely.

19. $|a_n| = 2/(n^3+e^n) < 2/n^3$, and $\sum |a_n|$ converges by the comparison test choosing $b_n = 2/n^3$ (p = 3). Thus it converges absolutely.

22. $|a_n| = \frac{|\cos n - 1|}{n^{3/2}} \le \frac{|\cos n| + 1}{n^{3/2}} \le \frac{2}{n^{3/2}}$. Since $\sum 2/n^{3/2}$ converges (p = 3/2), $\sum |a_n|$ converges by the comparison test. Thus the given series converges absolutely.

26. $\lim\limits_{n\to\infty} a_n = \lim\limits_{n\to\infty} \frac{\sec^{-1}n}{\tan^{-1}n} = \frac{\pi/2}{\pi/2} = 1$. Thus the series diverges.

28. Since $\lim\limits_{n\to\infty} |a_n| = \lim\limits_{n\to\infty} \frac{n^2+3}{(2n-5)^2} = \frac{1}{4} \ne 0$, $\lim\limits_{n\to\infty} a_n \ne 0$ and the series diverges.

31. $\lim\limits_{n\to\infty} \left|\frac{a_{n+1}}{a_n}\right| = \lim\limits_{n\to\infty} \frac{1\cdot3\cdot5\ldots(2n-1)(2n+1)}{(n+1)!} \cdot \frac{n!}{1\cdot3\cdot5\ldots(2n-1)} = \lim\limits_{n\to\infty} \frac{2n+1}{n+1} = 2 > 1$.

Thus the series diverges by the ratio test.

34. Since $a_n > 0$ for all n, we can drop the absolute value signs in the root test. $\lim\limits_{n\to\infty} \sqrt[n]{a_n} = \lim\limits_{n\to\infty} \frac{2n}{5n+3n^{-1}} = \frac{2}{5} < 1$. Thus the series converges by the root test.

37. Let $f(x) = x \tan(1/x)$, an $\infty \cdot 0$ indeterminate form as $x \to \infty$. $\lim\limits_{x\to\infty} f(x) =$

$\lim\limits_{x\to\infty} \frac{\tan(1/x)}{1/x} = \lim\limits_{x\to\infty} \frac{(-1/x^2)\sec^2(1/x)}{(-1/x^2)} = \sec^2 0 = 1$. Thus, $\lim\limits_{n\to\infty} a_n =$

$\lim\limits_{n\to\infty} n \tan(1/n) = 1 \neq 0$, and the series diverges.

EXERCISES 12.6, page 587

1. With $u_n = \dfrac{x^n}{n+4}$, $\lim\limits_{n\to\infty} \left|\dfrac{u_{n+1}}{u_n}\right| = \lim\limits_{n\to\infty} \left|\dfrac{x^{n+1}}{n+5}\cdot\dfrac{n+4}{x^n}\right| = \lim\limits_{n\to\infty} \left|\dfrac{n+4}{n+5}\right||x| = |x|$. So, we have absolute convergence by the ratio test if $|x| < 1$ or $-1 < x < 1$. If $x = 1$, the series is $\sum \dfrac{1}{n+4}$, which diverges. If $x = -1$, the series is $\sum \dfrac{(-1)^n}{n+4}$ which converges by the AST. Thus the interval of convergence is $[-1,1)$.

4. With $u_n = \dfrac{(-3)^n x^{n+1}}{n}$, $\lim\limits_{n\to\infty} \left|\dfrac{u_{n+1}}{u_n}\right| = \lim\limits_{n\to\infty} \dfrac{3n}{n+1} |x| = 3|x|$. So, we have absolute convergence if $3|x| < 1$ or $-\dfrac{1}{3} < x < \dfrac{1}{3}$. If $x = \dfrac{1}{3}$, the series is $\sum \dfrac{(-3)^n}{n3^{n+1}}$ $= \dfrac{1}{3}\sum \dfrac{(-1)^n}{n}$, which converges by the AST. If $x = -1/3$, the series is $\sum \dfrac{(-3)^n}{n(-3)^{n+1}} = -\dfrac{1}{3}\sum \dfrac{1}{n}$ which diverges. Thus the interval of convergence is $(-1/3, 1/3]$.

7. With $u_n = \dfrac{n}{n^2+1} x^n$, $\lim\limits_{n\to\infty} \left|\dfrac{u_{n+1}}{u_n}\right| = \lim\limits_{n\to\infty} \dfrac{n+1}{(n+1)^2+1}\cdot\dfrac{n^2+1}{n} |x| = |x|$. Thus, by the ratio test we have absolute convergence if $|x| < 1$ or $-1 < x < 1$. If $x = 1$, the series is $\sum \dfrac{n}{n^2+1}$ which diverges by the limit comparison test ($b_n = 1/n$), or the integral test. If $x = -1$, the series is $\sum \dfrac{n}{n^2+1} (-1)^n$, convergent by the AST. Thus the interval of convergence is $[-1,1)$.

10. $\lim\limits_{n\to\infty} \left|\dfrac{u_{n+1}}{u_n}\right| = \lim\limits_{n\to\infty} \dfrac{10^{n+2}}{3^{2(n+1)}}\cdot\dfrac{3^{2n}}{10^{n+1}} |x| = \lim\limits_{n\to\infty} \dfrac{10}{3^2} |x| = \dfrac{10}{9} |x|$. Thus we have absolute convergence if $\dfrac{10}{9} |x| < 1$, or $|x| < \dfrac{9}{10}$, or $-\dfrac{9}{10} < x < \dfrac{9}{10}$. If $x = \pm\dfrac{9}{10}$, $u_n = \dfrac{10^{n+1}}{3^{2n}} (\pm\dfrac{9}{10})^n$. Since $3^{2n} = (3^2)^n = 9^n$, $u_n = \dfrac{10^{n+1}}{9^n}\cdot\dfrac{9^n}{10^n}\cdot(\pm 1)^n$.

Thus, $u_n = 10(\pm 1)^n$, which does not have a limit as $n \to \infty$. (i.e., $u_n \not\to 0$ as $n \to \infty$.) Thus the power series diverges at each end point, and the interval of convergence is $(-\dfrac{9}{10} , \dfrac{9}{10})$.

13. $\lim\limits_{n\to\infty}\left|\dfrac{u_{n+1}}{u_n}\right| = \lim\limits_{n\to\infty}\dfrac{(n+1)!}{100^{n+1}}\cdot\dfrac{100^n}{n!}\,|x| = \lim\limits_{n\to\infty}\dfrac{(n+1)}{100}\,|x| = \infty$ unless $x = 0$, in which

case the limit is 0. Thus the series converges only for $x = 0$.

16. $\lim\limits_{n\to\infty}\left|\dfrac{u_{n+1}}{u_n}\right| = \lim\limits_{n\to\infty}\dfrac{\sqrt[3]{n}}{\sqrt[3]{n+1}}\,\dfrac{|x|}{3} = \dfrac{|x|}{3} < 1$ if $\dfrac{|x|}{3} < 1$ or $-3 < x < 3$. If $x = 3$, the

series is $\sum\dfrac{(-1)^{n-1}}{\sqrt[3]{n}}$ which converges by the AST. If $x = -3$, the series is

$\sum\dfrac{1}{\sqrt[3]{n}}$ which diverges $(p = 1/3)$. Thus the interval of convergence is $(-3,3]$.

19. $\lim\limits_{n\to\infty}\left|\dfrac{u_{n+1}}{u_n}\right| = \lim\limits_{n\to\infty}\dfrac{3^{2(n+1)}}{n+2}\cdot\dfrac{n+1}{3^{2n}}\,|x-2| = \lim\limits_{n\to\infty}\dfrac{9(n+1)}{n+2}\,|x-2| = 9|x-2| < 1$ if $|x-2|$

$< 1/9$ or $2 - \dfrac{1}{9} < x < 2 + \dfrac{1}{9}$. If $x = 2 + \dfrac{1}{9}$, $x-2 = \dfrac{1}{9}$, and the series is

$\sum\dfrac{1}{n+1}$, which diverges. (Remember $3^{2n} = (3^2)^n = 9^n$.) If $x = 2 - \dfrac{1}{9}$, $x-2 =$

$-1/9$ and the series is $\sum\dfrac{(-1)^n}{n+1}$ which converges by the AST. Thus the interval

of convergence is $[2 - \dfrac{1}{9}\,,\, 2 + \dfrac{1}{9}) = [\dfrac{17}{9}\,,\,\dfrac{19}{9})$.

22. $\lim\limits_{n\to\infty}\left|\dfrac{u_{n+1}}{u_n}\right| = \lim\limits_{n\to\infty}\dfrac{2n+1}{2n+3}\,|x+3| = |x+3| < 1$ if $-1 < x+3 < 1$ or $-4 < x < -2$. If

$x = -2$, $x+3 = 1$ and the series is $\sum\dfrac{1}{2n+1}$ which diverges. If $x = -4$, $x+3 = -1$

and the series is $\sum\dfrac{(-1)^n}{2n+1}$ which converges by the AST. Thus the interval of

convergence is $[-4,-2)$.

25. $\lim\limits_{n\to\infty}\left|\dfrac{u_{n+1}}{u_n}\right| = \lim\limits_{n\to\infty}\dfrac{n}{6(n+1)}\,|2x-1| = \dfrac{|2x-1|}{6} < 1$ if $|2x-1| < 6$ or $-6 < 2x-1 < 6$ or

$-\dfrac{5}{2} < x < \dfrac{7}{2}$. If $x = \dfrac{7}{2}$, $2x-1 = 6$, and the series is $\sum\dfrac{(-1)^n}{n}$ which converges

by the AST. If $x = -5/2$, $2x-1 = -6$, and the series is $\sum 1/n$ which diverges.

Thus the interval of convergence is $(-5/2,7/2]$.

28. $u_n = \dfrac{2\cdot4\cdot6\cdots(2n)}{4\cdot7\cdot10\cdots(3n+1)}\,x^n \Longrightarrow u_{n+1} = \dfrac{2\cdot4\cdot6\cdots(2n)(2(n+1))}{4\cdot7\cdot10\cdots(3n+1)(3(n+1)+1)}\,x^{n+1}$. Thus

$\left|\dfrac{u_{n+1}}{u_n}\right| = \dfrac{2\cdot4\cdot6\cdots(2n)(2n+2)}{4\cdot7\cdot10\cdots(3n+1)(3n+4)}\cdot\dfrac{4\cdot7\cdot10\cdots(3n+1)}{2\cdot4\cdot6\cdots(2n)}\,|x| = \dfrac{2n+2}{3n+4}\,|x|$, and

$\lim\limits_{n\to\infty}\left|\dfrac{u_{n+1}}{u_n}\right| = \lim\limits_{n\to\infty}\dfrac{2n+2}{3n+4}\,|x| = \dfrac{2}{3}\,|x|$. Thus, by the ratio test, the series con-

verges absolutely if $\dfrac{2}{3}\,|x| < 1$ or $|x| < \dfrac{3}{2}$, and the radius of convergence is $\dfrac{3}{2}$.

EXERCISES 12.7, page 593

1. $\frac{1}{1-x} = 1 + x + x^2 + \ldots = \sum_{n=0}^{\infty} x^n$ for $|x| < 1$. This is just the geometric

series encountered in Sec. 12.2, with $a = 1$ and $r = x$.

4. Replacing x by 4x in #1, $\frac{1}{1-4x} = \sum_{n=0}^{\infty} (4x)^n = \sum_{n=0}^{\infty} 4^n x^n$ for $|4x| < 1$ or $|x| < 1/4$.

7. $\frac{x}{2-3x} = \frac{x}{2}(\frac{1}{1 - \frac{3}{2}x}) = \frac{x}{2} \sum_{n=0}^{\infty} (\frac{3}{2}x)^n = \sum_{n=0}^{\infty} \frac{3^n x^{n+1}}{2^{n+1}}$ for $|\frac{3}{2}x| < 1$ or $|x| < \frac{2}{3}$, where

we replaced x by $\frac{3}{2}x$ in #1.

10. $\frac{3}{2x+5} = \frac{3}{5} \frac{1}{(1 + \frac{2}{5}x)} = \frac{3}{5} \sum_{n=0}^{\infty} (- \frac{2}{5}x)^n = \sum_{n=0}^{\infty} \frac{3(-1)^n 2^n x^n}{5^{n+1}}$ for $|\frac{2}{5}x| < 1$ or $|x| < \frac{5}{2}$,

where we replaced x by $(-\frac{2}{5}x)$ in #1.

13. Since $e^t = 1 + t + \frac{t^2}{2!} + \frac{t^3}{3!} + \ldots + \frac{t^n}{n!} + \ldots$ for all t, replacing t by -x we

obtain $e^{-x} = 1 + (-x) + \frac{(-x)^2}{2!} + \frac{(-x)^3}{3!} + \ldots + \frac{(-x)^n}{n!} + \ldots = 1 - x + \frac{x^2}{2!} - \frac{x^3}{3!}$

$+ \ldots + \frac{(-1)^n x^n}{n!} + \ldots$.

16. $\sinh x = \frac{1}{2}(e^x - e^{-x}) = \frac{1}{2}[(1 + x + \frac{x^2}{2!} + \frac{x^3}{3!} + \ldots) - (1 - x + \frac{x^2}{2!} - \frac{x^3}{3!} + \ldots)]$

$= \frac{1}{2}[2x + \frac{2x^3}{3!} + \frac{2x^5}{5!} + \ldots + \frac{2x^{2n-1}}{(2n-1)!} + \ldots] = x + \frac{x^3}{3!} + \frac{x^5}{5!} + \ldots +$

$\frac{x^{2n-1}}{(2n-1)!} + \ldots$.

19. Replacing x by $(-x^6)$ in #1, $\frac{1}{1+x^6} = 1 - x^6 + x^{12} - \ldots$ for $|x| < 1$. Thus

$\int_0^{1/3} \frac{1}{1+x^6} dx = [x - \frac{x^7}{7} + \frac{x^{13}}{13} - \ldots]_0^{1/3} = \frac{1}{3} - \frac{1}{3^7 \cdot 7} + \frac{1}{3^{13} \cdot 13} - \ldots$. This is a

convergent alternating series. Recall that if the sum of such a series is

approximated by the partial sum of the first n terms, the error is less than

the next term, i.e. the first term neglected. Since $\frac{1}{3^7 \cdot 7} \approx 6.532 \times 10^{-5}$ and

$\frac{1}{3^{13} \cdot 13} \approx 4.825 \times 10^{-7}$, we will have 6 place accuracy if we use the first two

terms. Thus the value is $\frac{1}{3} - \frac{1}{3^7 \cdot 7} \approx 0.33333 - 0.00007 = 0.33326$, which

becomes 0.3333 when rounded off to 4 places.

22. $\frac{x^3}{1+x^5} = x^3(1 - x^5 + x^{10} - \ldots)$ for $|x| < 1$. $\int_0^{.2} \frac{x^3}{1+x^5} dx = \int_0^{.2} (x^3 - x^8 +$

$$x^{13} - \ldots)dx = \frac{(.2)^4}{4} - \frac{(.2)^9}{9} + \ldots \ . \quad \text{Since the 2nd term is } \frac{512 \times 10^{-9}}{9} <$$

6×10^{-8}, we get 4 (in fact, 6) place accuracy with just the first term.

The value is $\frac{(.2)^4}{4}$ = .0004 to 4 places.

25. $f(x) = (1-x^2)^{-1} \Rightarrow f'(x) = 2x(1-x^2)^{-2}$. To obtain a series for $2x(1-x^2)^{-2}$,

we can start with the series for $f(x)$ and differentiate it. Replacing x by

x^2 in #1, we obtain $f(x) = \dfrac{1}{1-x^2} = 1 + x^2 + x^4 + x^6 + \ldots + x^{2n} + \ldots$, for

$|x| < 1$, whence $f'(x) = 2x(1-x^2)^{-2} = 2x + 4x^3 + 6x^5 + \ldots + 2nx^{2n-1}$, for

$|x| < 1$.

28. $\dfrac{e^t-1}{t} = [(1 + t + \dfrac{t^2}{2!} + \dfrac{t^3}{3!} + \ldots + \dfrac{t^n}{n!} + \ldots)-1]/t = 1 + \dfrac{t}{2!} + \dfrac{t^2}{3!} + \ldots + \dfrac{t^{n-1}}{n!} +$

$\ldots \ . \quad \text{Thus } f(x) = \displaystyle\int_0^x \dfrac{e^t-1}{t}\,dt = \int_0^x (1 + \dfrac{t}{2!} + \dfrac{t^2}{3!} + \ldots + \dfrac{t^{n-1}}{n!} + \ldots)dt =$

$[t + \dfrac{t^2}{2(2!)} + \dfrac{t^3}{3(3!)} + \ldots + \dfrac{t^n}{n(n!)} + \ldots]\Big|_0^x = x + \dfrac{x^2}{2(2!)} + \dfrac{x^3}{3(3!)} + \ldots + \dfrac{x^n}{n(n!)} + \ldots$

$= \displaystyle\sum_{n=1}^{\infty} \dfrac{x^n}{n(n!)} \ .$

EXERCISES 12.8, page 600

1. We calculate: $f(x) = \cos x$ $f(0) = 1$

                      $f'(x) = -\sin x$ $f'(0) = 0$

                      $f''(x) = -\cos x$ $f''(0) = -1$

                      $f'''(x) = \sin x$ $f'''(0) = 0$

and the subsequent derivatives follow this pattern. Thus $\cos x =$

$(1 - \dfrac{1}{2!} x^2 + \dfrac{1}{4!} x^4 - \dfrac{1}{6!} x^6 + \ldots) = \displaystyle\sum_{n=0}^{\infty} \dfrac{(-1)^n x^{2n}}{(2n)!} \ . \quad \text{The remainder, } R_n(x),$

is $(\pm \sin z)x^{n+1}/(n+1)!$ if n is even or $(\pm \cos z)x^{n+1}/(n+1)!$ if n is odd. In

either case $|R_n(x)| \leq \dfrac{x^{n+1}}{(n+1)!}$ which has limit 0 as $n \to \infty$ for every x by (12.42).

4. All even-numbered derivatives of $f(x) = \cosh x$ are $\cosh x$ and have value 1 at

x = 0. All odd-numbered derivatives are $\sinh x$ which are all 0 at x = 0. Thus

$\cosh x = (1 + \dfrac{x^2}{2!} + \dfrac{x^4}{4!} + \ldots) = \displaystyle\sum_{n=0}^{\infty} \dfrac{x^{2n}}{(2n)!} \ , \quad \text{in agreement with (12.48).} \quad \text{Here,}$

$R_n(x) = \dfrac{\sinh z}{(n+1)!} x^{n+1}$ (if n is even) or $R_n(x) = \dfrac{\cosh z}{(n+1)!} x^{n+1}$ (if n is odd). In

either case $|R_n(x)| \leq \dfrac{e^z + e^{-z}}{2} \dfrac{|x|^{n+1}}{(n+1)!} \leq (\cosh z) \dfrac{|x|^{n+1}}{(n+1)!}$ since $|\sinh z| \leq$

cosh z. Thus $|R_n(x)| \leq (\cosh x) \frac{|x|^{n+1}}{(n+1)!}$ since z between 0 and x $\Rightarrow$ cosh z $\leq$

cosh x. Again by (12.42), the factor $\frac{|x|^{n+1}}{(n+1)!} \to 0$ as $n \to \infty$ and $|R_n(x)| \to 0$ for all x.

7. Using $e^x = (1 + x + \frac{x^2}{2!} + \ldots)$, $e^{-x} = (1 - x + \frac{x^2}{2!} - \ldots)$, and sinh x =

$(e^x - e^{-x})/2$ we obtain sinh x = $x + \frac{x^3}{3!} + \frac{x^5}{5!} + \ldots = \sum_{n=0}^{\infty} \frac{x^{2n+1}}{(2n+1)!}$. With $u_n =$

$\frac{x^{2n+1}}{(2n+1)!}$, $\lim_{n\to\infty} \left| \frac{u_{n+1}}{u_n} \right| = \lim_{n\to\infty} \frac{x^2}{(2n+3)(2n+2)} = 0$ for all x. Hence r = ∞.

10. Substituting x^2 for x in #1, we obtain $\cos(x^2) = \sum_{n=0}^{\infty} \frac{(-1)^n (x^2)^{2n}}{(2n)!} =$

$\sum_{n=0}^{\infty} \frac{(-1)^n x^{4n}}{(2n)!}$. The ratio test yields $\lim_{n\to\infty} \left| \frac{u_{n+1}}{u_n} \right| = \lim_{n\to\infty} \frac{|x|^4}{(2n+2)(2n+1)} = 0$ for

all x. Hence r = ∞.

13. We calculate: f(x) = sin x $f(\pi/4) = 1/\sqrt{2}$

$\qquad\qquad\qquad$ f'(x) = cos x $f'(\pi/4) = 1/\sqrt{2}$

$\qquad\qquad\qquad$ f''(x) = -sin x $f''(\pi/4) = -1/\sqrt{2}$

$\qquad\qquad\qquad$ f'''(x) = -cos x $f'''(\pi/4) = -1/\sqrt{2}$,

and the higher derivatives follow the same pattern. Thus, the desired

Taylor's series is sin x = $\frac{1}{\sqrt{2}} + \frac{1}{\sqrt{2}}(x - \frac{\pi}{4}) - \frac{1}{2!\sqrt{2}}(x - \frac{\pi}{4})^2 - \frac{1}{3!\sqrt{2}}(x - \frac{\pi}{4})^3 + \ldots$

(Note that the pattern of signs is + + - - + + - -, etc.)

16. $f(x) = e^x \Rightarrow f^{(n)}(x) = e^x$ and $f^{(n)}(-3) = e^{-3}$ for all integers n $\geq$ 0. Thus

$e^x = \sum_{n=0}^{\infty} \frac{e^{-3}}{n!} (x-(-3))^n = \sum_{n=0}^{\infty} \frac{e^{-3}}{n!} (x+3)^n$.

19. Since powers of (x+1) are required, we want the Taylor series about c = -1. Thus

$\qquad\qquad$ $f(x) = e^{2x}$ $\qquad\qquad\qquad\qquad$ $f(-1) = e^{-2}$

$\qquad\qquad$ $f'(x) = 2e^{2x}$ $\qquad\qquad\qquad\qquad$ $f'(-1) = 2e^{-2}$

$\qquad\qquad$ $f^{(n)}(x) = 2^n e^{2x}$ $\qquad\qquad\qquad$ $f^{(n)}(-1) = 2^n e^{-2}$,

and $e^{2x} = \sum_{n=0}^{\infty} \frac{2^n e^{-2}}{n!} (x+1)^n$.

22. $f(x) = \tan x$ $f(\pi/4) = 1$

$f'(x) = \sec^2 x$ $f'(\pi/4) = 2$

$f''(x) = 2\sec^2 x \tan x$ $f''(\pi/4) = 4$

$f'''(x) = 2\sec^4 x + 4\sec^2 x \tan^2 x$ $f'''(\pi/4) = 16$,

and the first four terms are

$$1 + 2(x - \tfrac{\pi}{4}) + \tfrac{4}{2!}(x - \tfrac{\pi}{4})^2 + \tfrac{16}{3!}(x - \tfrac{\pi}{4})^3 + \ldots$$

25. $f(x) = xe^x$ $f(-1) = -e^{-1}$

$f'(x) = (x+1)e^x$ $f'(-1) = 0$

$f''(x) = (x+2)e^x$ $f''(-1) = e^{-1}$

$f'''(x) = (x+3)e^x$ $f'''(-1) = 2e^{-1}$

$f^{(4)}(x) = (x+4)e^x$ $f^{(4)}(-1) = 3e^{-1}$,

and the first four terms of the Taylor series are

$$xe^x = -e^{-1} + \frac{e^{-1}}{2!}(x+1)^2 + \frac{2e^{-1}}{3!}(x+1)^3 + \frac{3e^{-1}}{4!}(x+1)^4 + \ldots$$

28. Using the Maclaurin series for sin x from Example 1 and the fact that $1^\circ =$ $\pi/180$ radians, we obtain: $\sin 1^\circ = \frac{\pi}{180} - \frac{1}{3!}(\frac{\pi}{180})^3 + \frac{1}{5!}(\frac{\pi}{180})^5 - \ldots$. If we use only the first term to approximate the desired sum, the error is less than the next term since the series satisfies the conditions of the AST. Since $\frac{1}{3!}(\frac{\pi}{180})^3 \approx \frac{(0.01745)^3}{6} \approx 8 \times 10^{-7}$, we obtain $\sin 1^\circ \approx \frac{\pi}{180} \approx 0.01745$ to 5 places (0.0175 to 4 places).

31. $\displaystyle\int_0^1 \frac{1-\cos x}{x^2}\,dx = \int_0^1 \frac{1 - (1 - x^2/2! + x^4/4! - x^6/6! + x^8/8! - \ldots)}{x^2}\,dx =$

$\displaystyle\int_0^1 (\frac{1}{2!} - \frac{x^2}{4!} + \frac{x^4}{6!} - \frac{x^6}{8!} + \ldots)\,dx = \frac{x}{2} - \frac{x^3}{3(4!)} + \frac{x^5}{5(6!)} - \frac{x^7}{7(8!)} + \ldots\Big]_0^1 =$

$\frac{1}{2} - \frac{1}{72} + \frac{1}{3600} - \frac{1}{7(40,320)} + \ldots$. The fourth term is $< 4 \times 10^{-6}$, and, thus, the first 3 terms are sufficient for 5 place accuracy. The value of the integral, then, is $0.50000 - 0.01389 + 0.00028 = 0.48639 \approx 0.4864$.

34. $\cos t = 1 - \frac{t^2}{2!} + \frac{t^4}{4!} - \ldots \implies \cos x^2 = 1 - \frac{x^4}{2!} + \frac{x^8}{4!} - \ldots$

$\displaystyle\int_0^{.5} \cos x^2\,dx = \int_0^{.5} (1 - \frac{x^4}{2} + \frac{x^8}{24} - \ldots)\,dx = 0.5 - \frac{(0.5)^5}{10} + \frac{(0.5)^9}{9(24)} - \ldots$.

The 3rd term $= 1/2^9(216) = 1/(512)(216) < 1/(500)(200) = 10^{-5}$. Thus the first two terms are sufficient, and the value is $0.5000 - 0.00312 = 0.49688$ ≈ 0.4969.

EXERCISES 12.9, page 603

1. (a) Using (12.52) with k = 1/2, $\sqrt{1+x} = (1+x)^{1/2} = 1 + \frac{1}{2}x + \frac{(1/2)(-1/2)}{2!}x^2$

$+ \frac{1/2(-1/2)(-3/2)}{3!}x^3 + \ldots + \frac{(1/2)(-1/2)\ldots(1/2-n+1)}{n!}x^n + \ldots$ which reduces

to the answer given. By (12.58), r = 1.

(b) Using part (a) and replacing x by $-x^3$, we have $\sqrt{1-x^3} = 1 - \frac{x^3}{2} +$

$\sum_{n=2}^{\infty} (-1)^{n-1} \frac{1 \cdot 3 \ldots (2n-3)}{2^n n!} (-x^3)^n$ which again reduces to the answer given.

4. Using (12.52) with k = -2, $x(1+2x)^{-2} = x[1 + (-2)(2x) + \frac{(-2)(-3)(2x)^2}{2!} + \ldots$

$+ \frac{(-2)(-3)\ldots(-2-n+1)(2x)^n}{n!} + \ldots] = x[1 + \sum_{n=1}^{\infty} (-1)^n \frac{2 \cdot 3 \ldots (n+1) 2^n}{n!} x^n]$. By

(12.58), we have convergence if $|2x| < 1$ or $|x| < 1/2$. Thus r = 1/2.

7. We begin by computing $(1+x)^{-1/2} = 1 - \frac{1}{2}x + \frac{(-1/2)(-3/2)}{2!}x^2 + \ldots +$

$\frac{(-1/2)(-3/2)\ldots(-1/2 - n + 1)}{n!}x^n + \ldots = 1 + \sum_{n=1}^{\infty} (-1)^n \frac{1 \cdot 3 \cdot 5 \ldots (2n-1)}{2^n n!} x^n$.

Setting $x = -t^2$, we have $\frac{1}{\sqrt{1-t^2}} = 1 + \sum_{n=1}^{\infty} (-1)^n \frac{1 \cdot 3 \cdot 5 \ldots (2n-1)}{2^n n!} (-1)^n t^{2n}$.

Using $(-1)^n \cdot (-1)^n = (-1)^{2n} = 1$, we obtain $\sin^{-1}x =$

$\int_0^x \frac{1}{\sqrt{1-t^2}} dt = x + \sum_{n=1}^{\infty} \frac{1 \cdot 3 \cdot 5 \ldots (2n-1)x^{2n+1}}{(2n+1) 2^n n!}$. r = 1 since the series expansion

was valid for $|x| = |t^2| < 1 \iff |t| < 1$, and integrating does not alter r

by (12.42).

9. (This will be done rather than #10 since it relates to #1(b) above.)

By #1(b), $\sqrt{1-x^3} = 1 - \frac{x^3}{2} - \frac{x^6}{8} - \frac{3x^9}{48} + \ldots$ and, replacing x by -x, we have

$\sqrt{1+x^3} = 1 + \frac{x^3}{2} - \frac{x^6}{8} + \frac{x^9}{16} - \ldots$ and $\int_0^{1/2} \sqrt{1+x^3} dx = [x + \frac{x^4}{8} - \frac{x^7}{56} + \frac{x^{10}}{160} - \ldots]_0^{1/2}$

$= [\frac{1}{2} + \frac{1}{2^4 \cdot 8} - \frac{1}{2^7 \cdot 56} + \frac{1}{2^{10} \cdot 160} - \ldots]$. Since this is an alternating series,

the error in using only the 1st 2 terms is less than the 3rd term. Since

$\frac{1}{2^7 \cdot 56} = \frac{1}{7168} \approx 0.00014$, this will yield 3 place accuracy. Thus the value of

the integral to 3 places is $\frac{1}{2} + \frac{1}{2^4 \cdot 8} = \frac{1}{2} + \frac{1}{128} = 0.5000 + 0.0078 \approx 0.508$.

EXERCISES 12.10 (Review), page 605

1. $f(x) = \frac{\ln(x^2+1)}{x} \implies \lim_{x \to \infty} f(x) = \lim_{x \to \infty} \frac{\ln(x^2+1)}{x} = \lim_{x \to \infty} \frac{2x/(x^2+1)}{1} = 0$ by

L'Hôpital's rule. Thus $\lim\limits_{n\to\infty} \dfrac{\ln(n^2+1)}{n} = 0$ also.

4. If n is even, $(-2)^n$ is positive and becomes larger as n does. If n is odd, $(-2)^n$ is negative and becomes more so as n increases. So, even though $1/n \to 0$, the sequence behaves essentially like $(-2)^n$ which oscillates wildly, and no limit exists.

7. A preliminary analysis suggests that the terms behave like $1/\sqrt[3]{n^3} = 1/n$. Thus with $a_n = 1/\sqrt[3]{n(n+1)(n+2)}$ and $b_n = 1/n$, $\lim\limits_{n\to\infty} \dfrac{a_n}{b_n} = \lim\limits_{n\to\infty} \dfrac{n}{\sqrt[3]{n(n+1)(n+2)}} = 1$. Since $\sum b_n$ diverges (p = 1, the harmonic series), the given series diverges also.

10. $\lim\limits_{n\to\infty} \dfrac{1}{2+(1/2)^n} = \dfrac{1}{2} \neq 0 \Rightarrow$ the series diverges.

13. Since $\dfrac{n!}{\ln(n+1)} \geq \dfrac{n}{\ln(n+1)}$ for all $n \geq 1$, and since $\lim\limits_{n\to\infty} \dfrac{n}{\ln(n+1)} = \lim\limits_{n\to\infty} \dfrac{1}{1/(n+1)}$ $= \infty$ (by L'Hôpital's rule), the series diverges.

16. A preliminary analysis suggests we choose $b_n = 1/n^2$. So, with $a_n = \dfrac{n + \cos n}{n^3 + 1}$, $\lim\limits_{n\to\infty} \dfrac{a_n}{b_n} = \lim\limits_{n\to\infty} \dfrac{n^3 + n^2\cos n}{n^3 + 1} = \lim\limits_{n\to\infty} \dfrac{1 + (\cos n)/n}{1 + 1/n^3} = 1$. Since $\sum b_n$ converges (p = 2), the given series converges.

19. The series diverges since $\lim\limits_{n\to\infty} \sqrt[n]{n} = 1$, and thus, $\lim\limits_{n\to\infty} 1/\sqrt[n]{n} = 1 \neq 0$. To obtain this limit, let $y = x^{1/x}$, an ∞^0 indeterminate form as $x \to \infty$. Then $\ln y = \dfrac{1}{x} \ln x$, and $\lim\limits_{x\to\infty} \ln y = \lim\limits_{x\to\infty} \dfrac{\ln x}{x} = \lim\limits_{x\to\infty} \dfrac{1/x}{1} = 0$. Thus $\lim\limits_{x\to\infty} y = \lim\limits_{x\to\infty} x^{1/x} = e^0 = 1$, and $\lim\limits_{n\to\infty} n^{1/n} = \lim\limits_{n\to\infty} \sqrt[n]{n} = 1$ also.

22. If u_n denotes the nth term of the series, then $|u_n| = \dfrac{\sqrt[3]{n-1}}{n^2-1} \approx \dfrac{\sqrt[3]{n}}{n^2} = \dfrac{1}{n^{5/3}}$.

    This preliminary analysis suggests we use $b_n = 1/n^{5/3}$ in the limit comparison test. $\lim\limits_{n\to\infty} \dfrac{|u_n|}{b_n} = \lim\limits_{n\to\infty} \dfrac{n^{5/3}\sqrt[3]{n-1}}{n^2-1} = \lim\limits_{n\to\infty} \dfrac{\sqrt[3]{1-1/n}}{1-1/n^2} = 1$, having divided numerator and denominator by n^2. Since $\sum b_n$ converges (p = 5/3), so does $\sum |u_n|$, and the given series converges absolutely.

25. $\left|\dfrac{1 - \cos n}{n^2}\right| = \dfrac{|1 - \cos n|}{n^2} \leq \dfrac{1 + |\cos n|}{n^2} \leq \dfrac{2}{n^2}$. Since $\sum \dfrac{2}{n^2}$ converges (p = 2), the given series converges absolutely.

28. $\lim\limits_{n\to\infty}\left|\dfrac{u_{n+1}}{u_n}\right| = \lim\limits_{n\to\infty}\dfrac{3^n}{(n+1)^2+9}\cdot\dfrac{n^2+9}{3^{n-1}} = \lim\limits_{n\to\infty}3\,\dfrac{n^2+9}{(n+1)^2+9} = 3(1) = 3 > 1.$ Thus the

series diverges by the ratio test.

31. With $f(x) = \sqrt{\ln x}/x$ we have $f'(x) = (1-2\ln x)/2x^2\sqrt{\ln x} < 0$ for $x \geq 2$.

Moreover, $\lim\limits_{x\to\infty} f(x) = \lim\limits_{x\to\infty}\dfrac{\sqrt{\ln x}}{x} = \lim\limits_{x\to\infty}\dfrac{1/2x\sqrt{\ln x}}{1} = 0.$ Thus, the given series,

$\sum(-1)^n\,\dfrac{\sqrt{\ln n}}{n}$ converges by the AST. Now, if $n \geq 3$, $\dfrac{\sqrt{\ln n}}{n} > \dfrac{1}{n}$ ($\ln x$ is an in-

increasing function $\Rightarrow$ $\ln n \geq \ln 3 > \ln e = 1$). Since $\sum\dfrac{1}{n}$ diverges (harmonic

series), $\sum\left|(-1)^n\,\dfrac{\sqrt{\ln n}}{n}\right| = \sum\dfrac{\sqrt{\ln n}}{n}$ diverges by the comparison test. Thus the

original series converges conditionally.

34. $f(x) = \dfrac{x}{\sqrt{x^2-1}}$ is continuous, positive and decreasing on $[2,\infty)$. ($f'(x) =$

$-1/(x^2-1)^{3/2} < 0.$) $\displaystyle\int_2^\infty f(x)\,dx = \lim\limits_{t\to\infty}\int_2^t x(x^2-1)^{-1/2}\,dx = \lim\limits_{t\to\infty}(x^2-1)^{1/2}\Big]_2^t =$

$\lim\limits_{t\to\infty}(\sqrt{t^2-1} - \sqrt{3}) = \infty$, and the series diverges. (Note: divergence can be

obtained immediately by observing $n/\sqrt{n^2-1} \to 1$ as $n \to \infty$.)

37. $f(x) = 10/\sqrt[3]{x+8}$ is positive, continuous and decreasing on $[1,\infty)$. $\displaystyle\int_1^\infty f(x)\,dx =$

$\lim\limits_{t\to\infty}\displaystyle\int_1^t 10(x+8)^{-1/3}\,dx = \lim\limits_{t\to\infty}(15(t+8)^{2/3}-15(9)^{2/3}) = \infty$, and the series

diverges.

40. The series converges by the AST. Since the 10th term, $\dfrac{1}{100(101)} < 10^{-4}$, we

will get at least 3 place accuracy if we take only the 1st 9 terms as the

approximation to the sum, A. Thus $A \approx \dfrac{1}{1(2)} - \dfrac{1}{4(5)} + \dfrac{1}{9(10)} - \dfrac{1}{16(17)} + \dfrac{1}{25(26)}$

$- \dfrac{1}{36(37)} + \dfrac{1}{49(50)} - \dfrac{1}{64(65)} + \dfrac{1}{81(82)}.$ $A \approx .5000 - .05000 + .0111 - .0037 +$

$.0015 - .0008 + .0004 - .0002 + .0002 \approx .4585 \approx .458.$

43. With $u_n = \dfrac{(x+10)^n}{n2^n}$, $\lim\limits_{n\to\infty}\left|\dfrac{u_{n+1}}{u_n}\right| = \lim\limits_{n\to\infty}\dfrac{n}{2(n+1)}\,|x+10| = \dfrac{|x+10|}{2} < 1$ if $|x+10| <$

2 or $-2 < x+10 < 2$ or $-12 < x < -8$. If $x = -8$, $x+10 = 2$, and the series is

$\sum\dfrac{1}{n}$ which diverges. If $x = -12$, the series is $\sum\dfrac{(-1)^n}{n}$ which converges by the

AST. Thus the interval of convergence is $[-12,8)$.

46. With $u_n = \dfrac{(x+5)^n}{(n+5)!}$, $\lim\limits_{n\to\infty}\left|\dfrac{u_{n+1}}{u_n}\right| = \lim\limits_{n\to\infty}\dfrac{|x+5|}{n+6} = 0$ for all x. Thus the interval

of convergence is $(-\infty, \infty)$.

49. The quickest way for this problem is to recognize that $\sin x \cos x = \frac{1}{2}\sin 2x$

$$= \frac{1}{2} \sum_{n=0}^{\infty} \frac{(-1)^n (2x)^{2n+1}}{(2n+1)!} = \sum_{n=0}^{\infty} \frac{(-1)^n 2^{2n} x^{2n+1}}{(2n+1)!} \ . \quad r = \infty \text{ by the ratio test or from}$$

the nature of the sine series. Another method for this problem would be to multiply the Maclaurin series for $\sin x$ and $\cos x$ together.

52. See the solution to #7, Sec. 12.9.

55. One way is to write $\sqrt{x} = \sqrt{4+(x-4)} = 2\sqrt{1+(x-4)/4}$ and use #1(a), Sec. 12.9 replacing x by $(x-4)/4$ and multiplying by 2. Another way is to compute the Taylor series of $f(x) = x^{1/2}$ about $c = 4$. This yields $f(x) = x^{1/2}$, $f'(x) = \frac{1}{2}x^{-1/2}$, $f''(x) = \frac{1}{2}(-\frac{1}{2})x^{-3/2}$, $f'''(x) = \frac{1}{2}(-\frac{1}{2})(-\frac{3}{2})x^{-5/2}$, $\ldots$, $f^{(n)}(x) = (\frac{1}{2})(-\frac{1}{2})\ldots(\frac{-(2n-3)}{2}) \cdot x^{-(2n-1)/2}$ for $n \geq 2$. Evaluating these derivatives at $x = 4$ and using the Taylor series formula, the given answer results.

58. $\sin x = x - \frac{x^3}{6} + \frac{x^5}{120} - \frac{x^7}{5040} + \cdots$

$$\frac{\sin x}{\sqrt{x}} = x^{1/2} - \frac{x^{5/2}}{6} + \frac{x^{9/2}}{120} - \frac{x^{13/2}}{5040} + \cdots$$

$$\int_0^1 \frac{\sin x}{\sqrt{x}} \, dx = \frac{2}{3} x^{3/2} - \frac{2}{42} x^{7/2} + \frac{2}{1320} x^{11/2} -$$

$$\frac{2}{75,600} x^{15/2} + \cdots \ \Big]_0^1$$

$$= \frac{2}{3} - \frac{1}{21} + \frac{1}{660} - \frac{1}{37,800} + \cdots$$

Since the 4th term of this alternating series is $< 10^{-4}$, the sum of the 1st 3 yields 3 place accuracy at least. This sum is $.6667 - .0476 + .0015 = .6206$ $\approx .621$.